CAMBRIDGE LIBRARY COLLECTION

Books of enduring scholarly value

Latin American Studies

This series focuses on colonial Latin America and the Caribbean. It includes historical and statistical reference works from the eighteenth and nineteenth centuries, reports describing scientific, archaeological and ethnological expeditions, and editions of accounts from the earliest period of European settlement.

The History of a Voyage to the Malouine (or Falkland) Islands

After the French and Indian War, in which France and England had fought over the control of North America, the French government decided to send an expedition to the Falkland Islands, or Isles Malouines, with a view to founding a colony where the Acadians expelled from Canada could settle. Under the command of Louis-Antoine de Bougainville (1729–1811), the 1763–4 expedition claimed these islands for France and established a base at Port Saint Louis. This work, translated from French and first published in English in 1771, is an illustrated account of this endeavour by Antoine-Joseph Pernety (1716–96), a naturalist and writer who took part in the expedition. Notably, it contains the first description of the geological phenomena of the stone runs on East Falkland. The latter part of the work is a compilation of letters and notes pertaining to subsequent voyages to Patagonia and the Straits of Magellan.

The History of a Voyage to the Malouine (or Falkland) Islands

Made in 1763 and 1764,
under the Command of M. de Bougainville,
in Order to Form a Settlement There

Antoine-Joseph Pernety

Cambridge
University Press

CAMBRIDGE UNIVERSITY PRESS

Cambridge, New York, Melbourne, Madrid, Cape Town,
Singapore, São Paolo, Delhi, Mexico City

Published in the United States of America by Cambridge University Press, New York

www.cambridge.org
Information on this title: www.cambridge.org/9781108064330

This edition first published 1771
This digitally printed version 2013

ISBN 978-1-108-06433-0 Paperback

A Chart of the Atlantic and Southern Oceans.

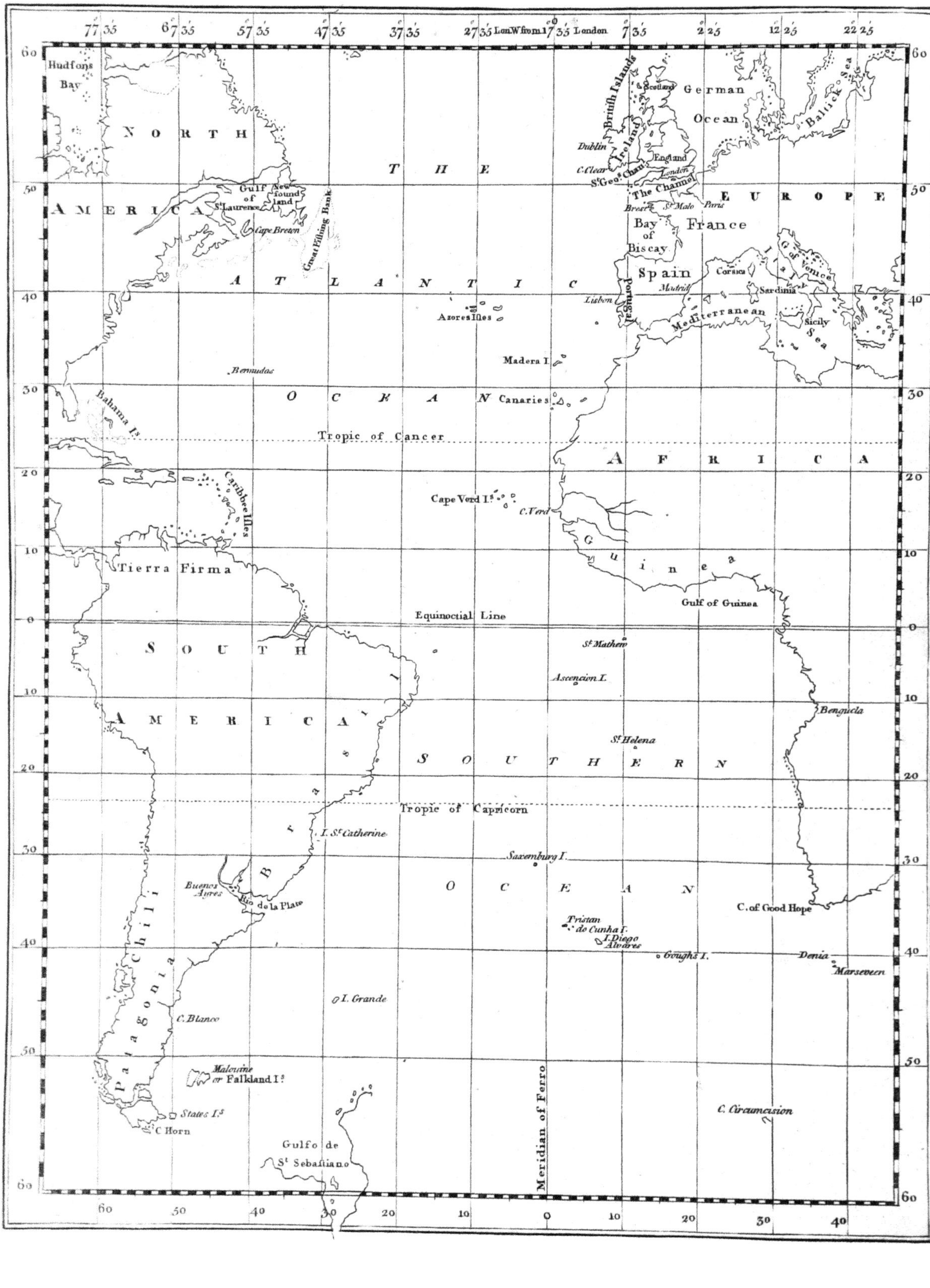

THE

HISTORY of a VOYAGE

TO THE

Malouine (or Falkland) Iſlands,

Made in 1763 and 1764,

Under the Command of M. de BOUGAINVILLE, in order to form a Settlement there;

AND OF

Two Voyages to the STREIGHTS of MAGELLAN,

WITH

An Account of the PATAGONIANS:

Tranſlated from Dom PERNETY's Hiſtorical Journal written in French.

ILLUSTRATED WITH COPPER PLATES.

LONDON:
Printed for T. JEFFERYS, in the Strand.
MDCCLXXI.

ADVERTISEMENT to the READER.

THE Editor thinks it neceſſary to acquaint the Reader, that, in this tranſlation of Dom Pernety's Journal, nothing has been omitted, but the detail of ordinary occurrences, which appeared common to every voyage. Whatever ſeemed in any view peculiar to this expedition has been retained.

In reſpect to the cuts and plans, ſome alterations and additions have been made. A general chart ſhewing the ſituation of Falkland's Iſlands in the Southern Ocean, which was not given in the original, is here inſerted. Plans of the iſlands of St. Catherine, and of Buenos Ayres, are alſo added. The birds, fiſh, &c. are claſſed in their proper order, and placed at the end of the book with references to the page in which they are mentioned.

The Editor hopes the work will meet with the approbation of the public, as he has ſpared neither coſt nor pains to make it uſeful and exact.

Note of the Tranſlator omitted in page 242.

The Pinguin here mentioned, is different from that which is deſcribed by our ingenious countryman Mr. Pennant, under the name of the Patagonian Pinguin; and anſwers more exactly to the ſecond ſpecies or leſſer Pinguin ſpoken of by that gentleman, and which is otherwiſe called, Anſer Magellanicus Cluſii, &c. It is probable that Don Pernety never ſaw the Patagonian Pinguin, ſince he ſays nothing of it. The Reader will find an accurate account of the different ſpecies of this ſingular bird, in the 58th volume of the Philoſophical Tranſactions, which contains Mr. Pennant's paper on that ſubject.

Directions to the Book Binder for placing the PLATES.

THE

AUTHOR's

PREFACE.

THE discovery and knowledge of the Malouine Islands has been looked upon as an object of so much importance, that the English, having been informed of the expedition we made there in 1764, thought it necessary to establish themselves in those islands, notwithstanding that we had already taken possession of them in the name of the crown of France. In preparing for this voyage, which excited the attention of all Europe, they took extraordinary precautions. Commodore Byron was employed on this expedition with two ships, the Dolphin and the Tamer frigate, under his command. The Florida was afterwards dispatched to carry them provisions of all kinds.

We had taken possession of these islands in the beginning of April, before the Dolphin was off the stocks, and we quitted them the 8th of the same month on our return to France, where we landed the 26th of June. The English did not sail till some days after. On the 4th of December they left Port Desire, and took their course towards the South of the supposed Pepys's Island, at 48 degrees South latitude, where they made several unsuccessful attempts for the discovery of that island. They were then obliged, as they observe p. 69. of the Voyage round the world in 1764 and 1765 on board the Dolphin, to abandon that research, being well persuaded of the impossibility of finding this supposed island.

The 22d of the same month (December) being in the Streights of Magellan, five leagues distance from Terra del Fuego, they observed a smoke rising in several places on the opposite coast, which is that of Patagonia. They steered towards it, and casting anchor at about a mile from ſhore, ſaw diſtinctly men on horſeback, who beckoned to them.

On approaching the coaſt there appeared manifeſt ſigns of fear in the countenances of thoſe, who were going on ſhore in the boat, when they perceived at the water ſide men of a prodigious ſtature. The Commodore, animated with the idea of making a diſcovery relative to theſe Patagonians, the queſtion of whoſe exiſtence had for a long time furniſhed matter of converſation in England, was the firſt who leaped on ſhore, and was followed by his officers and ſeamen well armed, whom he drew up in a poſture of defence. The ſavages, to the number of about 200, immediately ran up to them, looking at them with an air of the greateſt ſurprize, and ſmiling at the diſproportion in ſize between the Engliſh and themſelves.

The Commodore made ſigns to them to ſit down, which they complied with; and he put about their necks collars of enamelled beads, and ribbands, giving to each of them ſome ſuch trifling ornament. Their ſize is ſo extraordinary, that even ſitting they were almoſt as high as the Commodore when he ſtood upright. (p. 77.)

Their middle ſtature ſeemed to be about eight feet, and the higheſt above nine *. The Engliſh did not uſe any meaſure to aſcertain this; but we have reaſon, ſay they, to believe, the account we give rather falls ſhort of, than exceeds, the truth. (p. 78.) The ſize of the women is as ſurprizing as that of the men, and the children are in the ſame proportions. The women wore necklaces and bracelets. (p. 79.) Their cloaths were made of the ſkins of Peruvian ſheep, which covered their ſhoulders, and came down as far as their knees. The greateſt part of them were on horſeback, before we landed; but they alighted, and left their horſes at ſome diſtance. The horſes have the appearance of being

* The Commodore is ſaid, in the preface to the ſame Account, (p. 61.) to be ſix feet high. It muſt be remembered, that the Engliſh foot is near an inch leſs than the French ſtandard foot.

ing very ſwift, but their height is not in proportion to that of their riders, and they ſeemed beſides to be in but indifferent condition. (p. 85.) They appeared to be of a mild and friendly diſpoſition. (p. 83.)

Among the Engliſh was Lieutenant Cummins, whom the Patagonians ſeemed to regard with particular ſatisfaction, on account of his height, which was not leſs than ſix feet ten inches. Some of them clapped him on the back; but though this was intended only as a mark of their kindneſs, their hands fell ſo heavy upon him, that he ſtaggered under the weight of them.

On the 23d of the ſame month, the Engliſh having advanced farther into the Streights diſcovered ſeveral ſavages, on the Iſland of Saint Elizabeth, who made ſigns to them to come on ſhore. Both the men and women were of middling ſtature, and well ſhaped. Their hair was black; their ſkin, which is naturally of an olive colour, appeared red, becauſe they paint their bodies with a compoſition of reddiſh earth mixed with greaſe. They are cloathed with the ſkins of ſea-calves, otters, or Peruvian ſheep, ſewed together, ſo as to make one piece of about four feet and one half ſquare. They wear caps made of the ſkins of birds with the feathers, and have alſo ſkins on their feet, which ſerve them inſtead of ſhoes. Some of the women had girdles alſo made of ſkins; but none of them wore caps; they were only diſtinguiſhed by a necklace of ſhells. (p. 92.)

After having provided themſelves with wood and water at Port *Famine*, the Engliſh ſailed from thence the 5th of January 1765, and ſteering eaſtward, cleared the Streights, and ſaw land the 13th of the ſame month. The next day they entered a very commodious bay, within which were ſeveral ſmall ones, and different harbours: to the third of theſe they gave the name of Port Egmont. The entrance to this bay is by the North; it is half a mile in width, and has from ſeven to thirteen fathom depth on a muddy bottom. (p. 121).

The 23d of January, the Commodore took poſſeſſion of all theſe iſlands in the name of the King of Great Britain, and left them the 27th, without having eſtabliſhed any ſettlement there. Theſe iſlands are ſituated in 51 degrees 21 minutes South latitude, and 66 degrees ten minutes Weſt longitude. (p. 134). From hence they returned, coaſting along to the Streights of Magellan.

It will appear by the particulars of this Engliſh account, and by thoſe of my Journal, that we were acquainted with the Malouine Iſlands, and had formed a ſettlement there, near a twelvemonth before the two ſhips under the command of Mr. Byron had even diſcovered them. At the time even when theſe two veſſels arrived there, Monſ. de Bougainville was then returned; and having ſeen them from the port where he lay at anchor, ſet ſail for the Streights of Magellan, where he met with them, as will be ſeen at the end of my Journal.

I have entered into the detail of this Engliſh expedition to the Malouines in order to convince the public of the inconteſtable right of the crown of France to the poſſeſſion of them, in oppoſition to the injurious pretenſions of the Engliſh.

I have alſo given a ſketch of the account, which one of the officers of Mr. Byron's ſhip has printed concerning the giants of Patagonia, that the Reader might compare it with what is ſaid of them in the extracts from the journals of the French Captains, who have ſeen and made a longer ſtay with theſe Patagonians than the Engliſh have. Such a compariſon will prove to thoſe who are incredulous, or who have too much vanity to ſuffer themſelves to appear ignorant of what has never come to their knowledge, or, from the ſame principle, make a point of denying every thing they have not ſeen, that there exiſts, neverthelefs, a race of men, the bulk and enormity of whoſe ſize may teach theſe unbelieving, vain, and ſelf-conceited perſons, to reduce their magnificent pretenſions, and be contented to conſider themſelves as not the ſmalleſt among the race of dwarfs.

The

The Streights of Magellan were little known. The accounts we had of it till this time, though many in number, were not to be depended upon; the obſervations were either deficient in exactneſs or in perſpicuity. This has determined me to give thoſe of our two French Captains, and a chart of the Streights, corrected according to their obſervations.

It may be conjectured, and indeed with great appearance of probability, that the Malouine Iſlands formerly made a part of Patagonia and Terra del Fuego, and that they were ſeparated from them by violent earthquakes, which opened a paſſage for the ſea through a cleft cauſed by the eruption, and formed in time the channel, by which thoſe iſlands are divided from the continent. This conjecture is the more reaſonable, as the Terra del Fuego took its name from the volcanos, which were ſuppoſed to have been ſeen there, and as at ſome diſtance from that part of the Malouine Iſlands, where we have made our ſettlement, the hills and vallies ſhew clearly, by the diſorder of the beds of free ſtone, and the irregular heaps in which they lie, that this confuſion is the effect of earthquakes. See what is ſaid on this ſubject in my journal.

But what will aſtoniſh the Reader is, that a country ſo extenſive as the Malouines ſhould neither be inhabited by men, nor by any of thoſe quadrupeds, which are commonly met with among the Patagonians; and that the ſmall ſpider with long legs, which is called in France *Faucheuſe*, and the little brown cricket called *Cri-cri*, which is alſo found in chimneys, are the only two inſects we ſaw there. It is leſs wonderful, that we ſhould not meet with any of the reptile ſpecies, as travellers aſſure us, that there are none to be found in the territory of Chily, which lies to the Weſt of Patagonia, in the ſame parallel with the Malouine Iſlands.

Another motive, which induces me to believe, that the Malouines were originally joined to Patagonia, is, that there are no trees on them, and that the whole coaſt to the Eaſt of the Patagonians, and of Terra del Fuego, is without trees, to about 25 leagues

leagues up the country. At that diſtance ſome trees begin to appear, but from thence to the ſea ſide there is nothing to be found but ſhrubs and heath. It is the ſame on the Malouines. The diſcoveries which the Engliſh, who have ſettled more to the Weſt, may make on that ſide, will give us more light into theſe particulars. The Spaniards, who have ſucceeded my countrymen in the Eaſtern ſettlement will inform us with regard to thoſe parts.

The exactneſs of the plans and charts, as well as that of the figures of animals in the plates of my Journal, may be depended upon. The chart I give of the Rio de la Plata is the more intereſting as it was taken with the utmoſt accuracy, and as it is the only one of that river, the navigation of which is ſo dangerous.

INTRODUCTION.

AFTER the peace was concluded by a ceſſion of all Canada on the part of France to England, M. de Bougainville, Knight of St. Louis, and Colonel of infantry, conceived the deſign of indemnifying France for this loſs, if poſſible, by a diſcovery of the ſouthern continent, and of thoſe large iſlands, which lie in the way to it. A peruſal of admiral Anſon's voyage round the world fixed his ideas for finding the Malouine Iſlands, and determined him to make them the firſt object of his expedition, and to form a ſettlement there. He communicated his project to the miniſtry, who approved it. To carry it therefore into execution, M. de Bougainville cauſed a frigate and a ſloop to be built at St. Malo at his own expence, under the directions of the Sieurs Guyot du Clos and Chenart de la Gyraudais, who were to have the command of them under him. But being deſirous to make the execution of his deſign as advantageous as poſſible, and imagining that I might be of ſervice to him in that reſpect, he propoſed to me, juſt before he quitted Paris, to undertake the voyage with him. A few days after, I received the King's orders in a letter from the Duke de Choiſeul, miniſter for the marine department, to embark with M. de Bougainville. I made my diſpoſitions immediately for the voyage, and ſet off with him for St. Malo.

Thoſe, who are acquainted with the ſituation of the Malouine Iſlands, will applaud the project of M. de Bougainville; but few people have heard of thoſe iſlands, becauſe they were almoſt unknown. Some navigators had ſeen them, but, I think, I may aſſert, that no one before ourſelves had ever landed there, at leaſt in the part where we did. For this reaſon it will be proper to give ſome idea of the diſcovery of them from the accounts given by authors of eſtabliſhed reputation.

 Frezier,

Frezier, in his relation *du Voyage de la Mer du Sud*, printed in 4to, Paris 1716, p. 264, expreſſes himſelf thus: "If in this chart I have ſuppreſt ſome ſuppoſed countries, I have added others which are real, in the latitude of 51 degrees, and to which I have given the name of *new Iſlands*; becauſe they have been diſcovered ſince the year 1700, the greateſt part of them by the ſhips of St. Malo. I have placed them according to the reports of the Maurepas and St. Louis, ſhips belonging to the India Company, which had a near view of them, and the latter even took in freſh water there from a pond, which I have marked near Port St. Louis. The water here was reddiſh and ſomewhat inſipid, in other reſpects good for the ſea. Both theſe veſſels paſſed them in different parts, but the one which kept cloſeſt along the coaſt was the St. John Baptiſt, commanded by Doublet of Havre, who attempted to paſs through an opening he ſaw towards the middle of them; but perceiving ſeveral ſmall iſlands juſt riſing to the ſurface of the water, he thought proper to tack about. This cluſter of iſlands is the ſame which was diſcovered by Fouquet of St. Malo, and to which he gave the name of Anican, his owner. The routs I have traced will ſhew the bearing of theſe lands from the Streights of Le Maire, in her paſſage from which the St. John Baptiſt ſaw them, and from Statenland, which the two other ſhips had had a proſpect of before they found it.

The northern part of theſe lands, which is here called the *the Coaſts of the Aſſumption*, was diſcovered the 16th of July 1708, by Poré of St. Malo *, who named it after the ſhip he ſailed in. It was thought to be a new land, at the diſtance of about a hundred leagues eaſt of the new iſlands I am ſpeaking of; but I have

* It appears that Poré was not acquainted with the ſituation of the coaſts of the Patagonians, nor that of the new or Malouine Iſlands, or that he was miſtaken in his point. Theſe iſlands are in fact no more than 90 or 100 leagues diſtant from the Streights of Magellan; how then could they be at the diſtance of 100 leagues Weſt of the coaſt of the Aſſumption, as it is called by Pore? If he had known the ſituation of the Malouine Iſlands, he would have ſeen clearly by the latitude and longitude of the coaſt he ran along, that it could be no other than the coaſt of thoſe iſlands.

have made no ſcruple of joining it with them, having convincing reaſons to juſtify my opinion.

The firſt is, that the latitudes obſerved on the North and South of theſe iſlands, and the bearing of the known parts anſwer perfectly to the ſame point of reunion on the Eaſt ſide without leaving any ſpace between them. The ſecond, that there is no reaſon to imagine this coaſt lies eaſtward of the iſles of Anican. For M. Gobien of the St. John, who was pleaſed to communicate to me an extract of his journal, ſuppoſes it to lie South of the river Plata *; which account, taken ſtrictly, will not admit of its being at a greater diſtance than two or three degrees eaſtward, that is to ſay, five and twenty or thirty leagues †. But the difference of computations is always a mark of uncertainty. The firſt time they ſaw this coaſt on their paſſage from Saint Catherine's to the Braſils, they reckoned it at 329 degrees; the ſecond, in paſſing from the river Plata, where contrary winds obliged them to put in, after having tried to paſs Cape Horn: they ſuppoſed it at 322 degrees, and according to ſome 324; following the charts of Peter Goos, the errors of which we have taken notice of, ſo that little regard ought to be paid to them. However, as they relied upon them, they thought themſelves at a great diſtance from the Continent, and reckoning that they were too far eaſtward, ran three hundred leagues too far to Weſt in the South Sea, inſomuch that they imagined themſelves on the coaſt of Guinea, when they landed at Ylo. But the third and convincing

* The ſuppoſition of M. Gobien, of the St. John, is falſe, in placing this coaſt of Aſſumption South of the river Plata. We were on ſhore there, as he was, and in the ſame place, according to Frezier's chart, and found it by our computation about 64 degrees and a half W. longitude from the meridian of Paris, and the mouth of the river Plata 56° 30; which carries that part of the coaſt where M. Gobien and we landed eight degrees farther S. W. and anſwers nearly to the miſtake attributed by the author of Admiral Anſon's Voyage (p. 78.) to Frezier's chart in regard to the ſituation of the coaſt of Patagonia.

† If we place the coaſt of the Aſſumption three degrees farther to the Weſt, it will be more conformable to our eſtimation, which makes our landing place four degrees, or thereabouts, more to Weſt than it would be according to Frezier's chart, which is formed on the extract M. Gobien furniſhed him with from his own journal.

convincing argument is, that we ought to paſs to windward of this new land according to the longitude it was placed in, in the manuſcript chart; and that it is morally impoſſible any ſhip could paſs without ſeeing it, it being about 50 leagues in length from E. S. E. to W. N. W. No doubt therefore remains, that this was the northern part of the new iſlands, the weſtern part of which will be diſcovered in time, but is yet unknown.

Theſe iſlands are certainly the ſame, which were diſcovered by Sir Richard Hawkins in 1593, to the eaſt of the uninhabited coaſt, and in 50 degrees latitude. He was thrown by a ſtorm on an unknown land: he ran along the coaſt about ſixty leagues, and ſaw fires, from whence he concluded the place was inhabited *.

Hitherto theſe lands have been called Sebald's Iſlands, it being ſuppoſed that the three which go under this name in the charts were ſituated there at pleaſure, for want of a proper knowledge of them. But the ſhip *L'Incarnation*, commanded by the Sieur Brignon of St. Malo, took a near view of them in fine weather in the year 1711, on her departure from Rio Janeiro. They are in fact three ſmall iſlands † of about half a league in length, ranged

* I do not know whether the iſlands, which Sir Richard Hawkins ſaw in 1593, to the Eaſt of the deſart coaſt of Patagonia, in 50 degrees S. latitude, are the northern part of the new or Malouine Iſlands. We ran ſixty degrees at leaſt along the coaſt, as well as he, and ſaw no fire, or appearance of habitation, though we were very often at no greater diſtance than that of half a league or a league.

† At our landing we diſcovered three iſlands about half a league in length, pretty high, and forming a kind of triangle, agreeable to the account of Sebald's Iſlands. This reſemblance in poſition and figure made us take them at firſt for Sebald's Iſlands; but we diſcovered near them ſeveral ſmall flat iſlands, almoſt even with the ſurface of the water, of which no mention is made in the Sieur Brignon's journals, nor in thoſe of other people, who ſpeak of Sebald's Iſlands. A few hours after, having diſcovered other eminences, one behind another, we judged that theſe three iſlands were not Sebald's Iſlands, but ſome of the Malouines, which ſtand out before the principal one, and we found reaſon to confirm ourſelves in this opinion. If theſe three iſlands were really Sebald's Iſlands, they would be about two leagues diſtance from land, or the principal iſland, and not ſeven or eight, as Frezier ſays." See the chart of our route along the coaſt. However, in the two voyages of the Eagle and the Star Pink, which have taken a later view of theſe three iſlands in their paſſage from the Malouines to the Straits of Magellan, the Eagle in 1765, and the Eagle with the Star in 1766; theſe veſſels found no more iſlands than thoſe three, and have ſince looked upon them to be Sebald's Iſlands.

ranged in a triangular form, as they are repreſented in the charts. They paſſed at the diſtance of three or four leagues from them, and ſaw no land, though the weather was very fine, which is a proof that they are ſeparated from the new iſlands by at leaſt ſeven or eight leagues.

In the memorial preſented to the *Compagnie des Indes* by the Sieur de Lozier Bouvet in the year 1735, ſoliciting their aſſiſtance in furniſhing him the means of obſerving the countries diſcovered by Gonneville, he reports, among other advantages of the eſtabliſhment they might form there after that obſervation, the opportunities of fixing an immediate commerce with the Spaniards of the river Plata and the Portugueſe of Braſil. He even aſſerts, that the ſhips, in putting into the ſouthern coaſts, would ſteer very little out of their ordinary courſe for India.

By the eſtabliſhment we have made on the Malouine Iſlands * we have put the India Company, and all the French navigators in the moſt favourable ſituation for accompliſhing theſe two objects. The Malouine Iſlands are not near ſo far to the South: the climate is much more temperate; they ſtand more convenient for the river Plata and the Braſils; more in the neighbourhood of Magellan's lands and Patagonia, with the inhabitants of which it would be ſo much the more eaſy to fix a commerce, as they are already acquainted with the Europeans by the traffic they carry on with the Spaniards.

Let us conſider the ſituation of the ſouthern lands diſcovered by Monſieur de Gonneville, a gentleman of Normandy. In 1503 he fitted out a veſſel at Honfleur, and ſet ſail in the month of June for the Eaſt Indies. After doubling the Cape of Good Hope, and meeting with a guſt of wind, which was ſucceeded by calms, he thought of nothing but gaining ſome land, where he might recover the fatigues of the voyage. He had the good fortune to diſcover ſome, and called them the South Indies. He lay there ſix months, during which time he refitted, formed an

* Since this journal was written, France has ceded the Malouine Iſlands to Spain.

intercourſe with the natives, and eſtabliſhed himſelf ſo far in their confidence, that their King, Aroſca, truſted his ſon Eſſomeric to him to make the voyage of France, on condition that he ſhould bring him back in twenty months. Gonneville ſailed from thence the third of July 1504 laden with the produce of the country. In the Channel he met with an Engliſh privateer, which took him, and carried him into Guernſey. This unlucky accident prevented his arriving in France till the year 1505; where he made his complaint and declaration to the admiralty at Honfleur. No advantage was made at that time of M. de Gonneville's diſcovery; who to make amends to Eſſomeric for not being able to keep his word with him, married him to one of his relations, and left him at his death half his fortune.

The Sieur Bouvet, who had ſome notion of this diſcovery, preſented a memorial to the *Compagnie des Indes*, who fitted out two ſhips for him, the Eagle, and Mary, with which he ſailed from l'Orient the 19th of July 1738. The 26th of November he got into 35 degrees South latitude and 344° longitude from the French meridian. Here he began to meet with fogs, which continued almoſt conſtantly while the two ſhips remained in company. They were often ſo thick, that the Eagle's crew could not diſcover the Mary at the diſtance of muſquet ſhot; ſo that they had the greateſt difficulty to keep together. The 3d of December, being in 39 degrees 20 minutes latitude, and 351 longitude, they began to diſcover ſome ſea-weed, and more birds than ordinary, which made them imagine they were not far from land: they therefore took all the precautions neceſſary in ſuch circumſtances. The 5th, they found themſelves in 42 degrees 40 minutes latitude, and 354° longitude. The 7th, in 44 latitude, and 355 longitude. The 10th, 44° latitude, and the firſt meridian, where ſeveral geographers place the neareſt point of the Southern Continent. The 12th, they made 7 degrees longitude; the 15th, 48 degrees, 50 minutes latitude, which is equal to that of Paris, in 7° longitude. Here they ſaw ice, which they looked upon as a certain indication of land. They even obſerved a

change

change in the colour of the ſea, and ſaw a great number of Puffins, and other birds, ſeveral of which flapped their wings, like land birds. They perceived Penguins alſo, an amphibious bird, a deſcription of which is given in my journal. In proportion as they got farther to the South, the ice increaſed. The 16th they ſaw Penguins again, and a ſea wolf; the fogs and ice prevented their riſing to the 54th degree of latitude before the laſt day of December. At length, on the firſt of January, about three in the afternoon, they diſcovered a high land, covered with ſnow, and very foggy, which they took for a large head-land, and called it *Cape Circumciſion.* It lies, according to the account of the Sieur Bouvet, in 54 degrees South latitude, and from 27 to 28 degrees longitude from the French meridian. The 6th, they ſaw a prodigious quantity of birds, of a very fine white, and of the ſize of pigeons: they thought they ſaw land at the diſtance of one or two leagues. The next day they perceived a new land, nearly North North Eaſt of Cape Circumciſion. They continued in ſearch of it till the 9th, at four in the morning, when the weather being fair and the fog gone off, they found that the ſuppoſed land was nothing more than a miſt.

From the time they came within ſight of land, they had reaped no other benefit from it than that of concluding, that it extended from eight to ten leagues E. N. E. and from ſix to ſeven leagues Eaſt. They had not been able to diſcover even, whether what they ſaw was an iſland, or whether it made part of the Continent. At length the bad weather came on, the ſeaſon was advanced, and the crew were in a bad ſtate of health. All theſe conſiderations induced M. de Lozier Bouvet to take the reſolution of going to ſee for ſome place to put in at, which might be more eaſy and more convenient for their landing. He took his courſe with a view of finding the place where Gonneville had landed; which, according to the account of it, is ſituated in a latitude equal to that of ſome of the provinces of France. The moſt northern lie in 51 degrees, which is the latitude of the Malouine Iſlands. He made therefore for the parallel from 51 to 52, and

paſſed

paſſed it with the ſame inconveniencies, without reaping any kind of advantage. They reckoned themſelves in 51° longitude, when they were really in 55°, as they found on landing at the Cape of Good Hope. They kept the Cape to the North of them, and continued that courſe till the 5th of February, when the two veſſels ſeparated, the Mary ſteering for the Cape with M. Bouvet, and the Eagle for the iſle of France with Mr. Hay.

M. Bouvet left the Cape of Good Hope the 31ſt of March, on his return to France, and in his route ſaw Trinity Iſland in 351 degrees longitude from the meridian of Teneriff, and 348° 30 from the French meridian, 20 degrees 20 minutes latitude. He likewiſe ſaw the iſle of Aſcenſion, which he places in 349 degrees longitude. He ſays, that Trinity Iſland is, without that name, very well deſcribed by the *Flambeau Anglois*. After we had got ſays M. Bouvet, within gun ſhot of this iſland, we ſaw three-fourths of it diſtinctly. It is properly ſpeaking, nothing more than a rock inacceſſible on all ſides. There are four little iſlands between 8 and 9 leagues Eaſt of it. Oliver de Noort, who had the command of four Dutch ſhips in 1599, followed this parallel of 20 degrees 20 minutes from this iſland as far as the coaſt of Brazil, and found no other in his courſe. This has made it imagined, that what is called Martin de Vaz's Iſland, and the Iſland of Aſcenſion are the ſame with Trinity Iſland *, which goes under theſe three different names. We have been more fortunate in our enterprize than M. Bouvet was in his. The ſettlement we made at the Malouine Iſlands would anſwer all the purpoſes of that he deſigned to make on the Southern Continent, if

* What M. Bouvet ſays here of Trinity Iſland is very conformable to what we ſaw near the iſland of Aſcenſion, which is recounted in this journal, on 27th April 1764. But though their ſituation in reſpect of latitude does not differ more than 12 minutes, the longitude is abſolutely different; ſince, according to his eſtimation, Trinity Iſland is at 348 degrees 30 minutes from the French meridian, which anſwers to about 10 degrees from the meridian of Paris. While we were reconnoitring the iſland of Aſcenſion, I eſtimated its ſituation at 32 degrees 25 minutes from the latter meridian. It ſhould follow from thence that Trinity Iſland and the Iſland of Aſcenſion are really diſtinct from each other; which is contrary to the opinion of ſeveral navigators.

if the India Company's ſhips would take their route by the South Sea to China, the Philippine Iſlands, &c. and for the South Sea trade. The author of Anſon's Voyage expreſſes himſelf upon this point in the following manner, page 54 & ſeq. 4to edition, printed for Charles Anthony Jombert. "I have proved above, that all our future expeditions to the South Seas muſt run a conſiderable riſque of proving abortive, whilſt we are under the neceſſity of touching at Brazil in our paſſage thither; an expedient therefore, that might relieve us from this difficulty, would ſurely be a ſubject worthy of the attention of the public." We may add, that this port is too far from the neareſt that can be found in the South Sea to be of ſufficient advantage. We put into St. Catherine's as well as Admiral Anſon: we had not indeed, like him, reaſon to complain of the reception we met with; on the contrary, we owe our acknowledgements to the Governor, as will be ſeen in this journal; but the other inconveniences of this harbour are ſuch as he has reported them. The unhealthy air and perpetual fogs, which are found there, are enough to create a diſguſt.

"The beſt method of effecting this, (ſays the ſame author) would without doubt be by a diſcovery of ſome place more to the ſouthward, where ſhips might refreſh, and ſupply themſelves with the neceſſary ſea-ſtock for their voyage round Cape Horn. And we have in reality the imperfect knowledge of two places, which might perhaps, on examination, prove extremely convenient for this purpoſe: the firſt of them is Pepys's Iſland * in the latitude of 47 degrees South, and laid down by Dr. Halley about eighty leagues to the eaſtward of Cape Blanco, on the coaſt of Patagonia; the ſecond is Falkland's Iſles in the latitude of 51° † nearly South of Pepys's Iſland. The firſt of theſe was diſcovered

* In the ſecond voyage to the Malouines M. de Bougainville endeavoured for ſeveral days, without ſucceſs, to find this ſuppoſed Pepys's Iſland: the ſame attempt was made in the third voyage, and proved equally unſucceſsful.

† *Note of the tranſlator.* The original Engliſh ſays 51° ½. The reſt, there being no very material difference, is copied verbatim from the original.

discovered by Captain Cowley in his voyage round the World, in the year 1686, who represents it as a commodious place for ships to wood and water at; and says, it is provided with a very good and capacious harbour, where a thousand sail of ships might ride at anchor in great safety; that it abounds with fowls, and as the shore is either rocks or sand, it seems to promise great plenty of fish."

This reasoning appears to be merely conjectural, and very boldly advanced on the part of Captain Cowley. It is easy to convince one's self of this by reading his relation, since he says in so many words, that *the bad weather hindered his landing there, he not having been able to put his longboat to sea*. If then he really did see it, it was only in his passage, as many navigators have a multitude of other islands and continents, which are still unknown to us, as well in respect to the quality and productions of the soil as to the real situation of their coasts. Since this captain did not go on shore there, how could he know, that it is a good place to water at? Perhaps there is no fresh water. As to wood, we have been deceived by appearances in running along the coast of the Malouines: we thought we saw some, and after landing, these appearances vanished into cornflags, a sort of rush or plant with long, flat, strait leaves, which grows on a hillock of three feet in heighth at least, and the leaves clustering together form, as they rise from the hillock, an eminence of six or seven feet. See the extract from the Sieur Alexander Guyot's journal at the end of this work.

"The second place, or Falkland's Isles, (proceeds the Admiral) have been seen by many ships both *French* and *English*, being the land laid down by Frezier in his chart of the extremity of South America under the title of the *new islands*. Woods Rogers, who ran along the N. E. coast of these isles in the year 1708, tells us, that they extended about two degrees in length, and appeared with gentle descents from hill to hill, and seemed to be good ground with woods and harbours (see what we have said in relation to this in the preceding paragraph). Either of

these

theſe places, as they are iſlands at a conſiderable diſtance from the Continent, may be ſuppoſed from their latitude, to lie in a climate ſufficiently temperate. It is true, they are too little known to be at preſent recommended for proper places of refreſhment for ſhips bound to the ſouthward: but if the admiralty ſhould think it adviſeable to order them to be ſurveyed, which may be done at a very ſmall expence, by a veſſel fitted out on purpoſe; and if, on this examination, one or both of theſe places ſhould appear proper for the purpoſe intended, it is ſcarcely to be conceived of what prodigious import a convenient ſtation might prove, ſituated ſo far to the ſouthward, and ſo near Cape Horn. The Duke and Ducheſs of Briſtol were but thirty-five days from their loſing ſight of Falkland's Iſles to their arrival at Juan Fernandez in the South Seas: and as the returning back is much facilitated by the weſtern winds I doubt not but a voyage might be made from Falkland's Iſles to Juan Fernandez, and back again in little more than two months."

If Woods Rogers only ran along the North Eaſt coaſt of Falkland's or the Malouine Iſles, how could he know, that they did not extend more than about two leagues? We ran along only one ſide of the coaſts of the principal iſland and found that it extended more than three degrees from Eaſt to North Eaſt. It is true, we obſerved, that it is compoſed of eminences with gentle deſcents from one to another, but the ground did never appear to us to be covered with wood, although we ſteered cloſe along the ſhore: we even doubted if there was any there, not having been able to find it during the ſtay we made in all the three voyages.

AN

HISTORICAL JOURNAL

OF MY

VOYAGE

TO THE

MALOUINE ISLANDS,

WITH THE

Obſervations I made on the Inhabitants and on the Natural Hiſtory of the Places I met with in my Way.

I LEFT Paris the 17th of Auguſt 1763, at two o'clock in the afternoon. We ſtopped at Pontchartrain, in expectation of M. d' Arboulin, at that time Adminiſtrator General of the poſts in France, who was returning from his eſtate at Montigny to Paris. M. de Bougainville, his nephew, was deſirous of conſulting with him on the arrangements neceſſary to be taken relative to the expences attending the building the two frigates, and the voyage we were going to undertake: M. d'Arboulin had a large ſhare in the undertaking. We waited for him till near ſeven o'clock; he came at laſt, and after a conference of about an hour ſet out for Paris, and we at the ſame time for St. Malo. We travelled the two following nights and days, ſtopping only at Rennes for a few hours in the middle of the day to let the heat go off, which was exceſſive, and to faſten one of our wheels, the ſpokes of which could not be kept in the nave. On Sunday, the 20th, about two in the morning, we arrived at Beauſejour. This is a very pretty country ſeat, ſituated at one end of St. Servant. M. Bougainville de Nerville, couſin-german to M. de Bougainville, had arrived there five days before us, and waited our coming. We drank but one glaſs of cyder and ran to bed, having more deſire to ſleep than eat.

M. Duclos Guyot, who had been pitched upon to command the Eagle frigate, under the orders of M. de Bougainville, came to meet us at Beauſejour, with ſome of the officers who were to embark with us. I paſſed my time in ſeeing the towns of St. Malo and St. Servant, and the environs, till the 25th, when we went to Port Solidor, for the ceremony of baptiſing our two frigates, which was performed with the uſual ſolemnities. All the officers and ſailors, who were to embark in them, were on board. M. N. chaplain and director of the hoſpital of St. Saviour, in the town of St. Malo, ſaid maſs on board the Eagle, and performed all the ceremonies cuſtomary upon ſuch occaſions. The two frigates, anchored cloſe by each other, gave a general ſalute at the beginning of the maſs, and another at the end during the prayer for the King.

The next day, Don Jamin, prior of the Benedictines of the convent of St. Benoit, with whom I had been much connected, while he was profeſſor of divinity in the abbey of St. Germain des Pres at Paris, entertained M. de Bougainville, Meſſ. Duclos Guyot, Chênart de la Gyraudais, de Belcourt, Lieutenant of infantry, l' Huillier de la Serre, Engineer, and myſelf, at dinner.

We embarked our baggage, beds, and other neceſſaries for the voyage, and the 29th we lay on board. It was the firſt of September before every thing was embarked.

By five o'clock that morning we left Solidor, with a briſk wind to N. W. in the frigate *Eagle*, with a crew of 100 men, mounting 20 guns, pierced for 24, commanded by the Sieur Duclos Guyot of St. Malo, Captain of a fireſhip, in company with the

ſloop

ſloop Sphinx, crew 40 men, mounting 8 guns and 6 ſwivels, commanded by the Sieur Chênart de la Gyraudais of St. Malo, Lieutenant of a frigate, both under the command of M. de Bougainville, Knight of St. Louis, Colonel of infantry, and Captain of a ſhip: at ten in the morning we were in the road of Rance, or St. Malo.

We were only waiting for a favourable wind to ſet ſail the next morning, the 2d of September, when three or four perſons of St. Malo raiſed difficulties at the Admiralty upon our departure. M. de Bougainville, having received notice of it, went on ſhore to St. Malo, appeared at the Admiralty, and anſwered every objection ſo fully, that judgment was given in his favour. Thinking, however, that it was proper to inform the miniſtry of this tranſaction, he ſent off a courier with diſpatches at two in the morning, Sunday the 4th. The courier, who was his own ſervant, made ſo much haſte, that he returned to St. Malo, with an anſwer, in fifty-nine hours from his ſetting off.

On the 8th at night, being the nativity of the Virgin, the wind appearing at S. S. W. orders were given to unmoor, which was accordingly done by one in the morning, and at half paſt ſix we ſet ſail, the gale continuing freſh.

We kept under ſail the 9th, and after having cleared the harbour, the wind being got about to S. W. and veering more and more to Weſtward, as we approached Cape Frehel, we came to anchor about noon. The ſloop Sphinx followed our example. Our two frigates were then in the ſame ſituation in which the Engliſh fleet was at the affair of St. Cas, where they were ſo roughly handled. This anchorage is by no means ſecure: many ſhips have been loſt here.

While we remained here, I took the opportunity of putting into a ſmall caſk, which held about ſix gallons of water, a compoſition of M. Sequin's, to preſerve water from ſpoiling in long voyages. A chymiſt had given another, for the ſame purpoſe, to M. de Bougainville. It was a paſte of a greyiſh caſt; which ſeemed to be made of clay, and the powder of crude antimony.

Some

Some ſaid, there was a mixture of crude mercury in it. But, as M. de Bougainville did not ſhew it to me before we got on board, I did not analyſe it; and he, under the uncertainty he was in with reſpect to the ingredients, was not very deſirous of trying the effects of it. For myſelf, as I knew the compoſition of M. Sequin's drug, which is nothing but ſpirit of ſalt, and that, at the ſame time, it preſerves the water from corruption, and renders it more wholeſome and uſeful in preventing or curing the ſcurvy, I made no ſcruple of trying it. What were the effects of it will be ſeen in the ſequel.

The 15th we got again under ſail, and the wind being ſtill contrary, got ſight of the light-houſe of Frehel the 17th, at the diſtance of about four leagues: on the 18th we reſolved to come to anchor. Accordingly we ſtood in to ſhore: the Sphinx did the ſame; and, after much difficulty in weathering the caſtle of la Latte, we anchored about two in the afternoon.

As the ſea was become very calm, and the wind tolerably quiet, by nine this morning, M. Bougainville, Meſſ. de Belcourt, l'Huillier, Donat, de la Gyraudais Captain of the Sphinx and myſelf, had been to the iſland *Agôt* to ſhoot rabbits; but we ſaw only two in the courſe of three hours. As I had no other game in view than the finding of plants, or other curioſities, that might happen to lye in my way, I amuſed myſelf with picking up the ſeeds of radiſhes, or wild horſe-radiſh, and ſome ſhells. Towards noon, we began to find ourſelves hungry, we had killed nothing, and were at a loſs for our dinner. Upon this we called a council, and it was reſolved to go and beg a dinner of the prior of St. Jacut. We went immediately into our boat, and got to the abbey about two o'clock. The prior, and the other Benedictines, my brethren, received us in the moſt obliging manner, and treated us with the ſame hoſpitality, which we had met with the ſixth of this month, when we dined five or ſix of us with the prior of Benedictines at St. Malo. The prior of St. Jacut had dined on board the Eagle the 13th, and M. de Bougainville had done the honours in the beſt manner.

As ſoon as dinner was over at St. Jacut, I put the prior in mind of the offer he had made us of greens from their garden. He, with great civility, gave us leave to take what we pleaſed, and we loaded our boat with cabbages and leeks.

On the 20th, at one in the afternoon, we ſhipped our boats, wind at S. S. W. briſk gale, inclinable to ſqualls. By three we were under ſail. After doubling the point of the caſtle of la Latte, we found the wind at N. W. ſo came to anchor again.

At nine in the evening we had a violent ſquall of wind, which laſted above half an hour. During this, an Acadian, one of our paſſengers, ſtood on the forecaſtle with his arms folded, and, while the crew were all employed, kept looking on with the utmoſt compoſure. M. de Bougainville, to whom a complaint had been made of this very man a few days before for the ſame kind of behaviour, and who had ſpoke to him about it, could not now refrain for giving him a reprimand. The Acadian, without returning an anſwer, went below deck, and there exclaimed loudly againſt this treatment to his wife, his father, and two other Acadian families, which were likewiſe paſſengers, adviſing them to follow his example; for after all, ſays he, we were not hired, nor taken on board to work our paſſage, but as volunteers and paſſengers; and, for my part, I would much rather have ſtaid in France, than have embarked on ſuch conditions.

All this diſcourſe was reported to M. de Bougainville, who was piqued at it, and with reaſon. Theſe Acadian families had lived at St. Servant, and St. Malo, ever ſince the Engliſh took Acadia from us. The King allowed them ſo much a head, in the ſame manner as his regular troops; and theſe families had ſcarce any other reſource than this ſort of pay and their own labour. M. de Bougainville offered to take them on board with him, and to carry them to a country where he would give them a landed property, and many other advantages, which they could never expect in France. He had even furniſhed them with goods and money in advance. Upon the report that was made to him of the diſcourſe of this Acadian, he ſaid, there was nothing more to be done than

to

to ſet them on ſhore, and ſend them back to St. Servant; ſince they were fond of miſery, they might go there and enjoy it. As ſoon as this was told to the other families, it made ſo great an impreſſion on them, that the women burſt into tears, and the men upbraided the Acadian, who had been the cauſe of it, and a diſagreement among them enſued. Of this M. de Bougainville was ſoon informed. The next day, the 21ſt, after prayers, he called them all before him; there are, ſaid he, ſome diſcontented perſons among you, who repent of having embarked with me. I do not require you to do the duty of common ſailors: I did not take you on board with me upon that footing; but, at the ſame time, I did not mean that you ſhould conſider yourſelves as mere paſſengers, and not lend a hand upon occaſion. You are at liberty to go back to St. Malo, St. Servant, or whatever place you think fit; you have only to ſpeak, and you will be ſet on ſhore immediately.

The Acadian and his father declared, they choſe to return to St. Servant. The two other families deſired to go the voyage. Early in the afternoon the father, the ſon and his wife were landed near St. Caſt, with their effects; and M. de Bougainville, out of charity, left them the money he had obtained in advance for them from the King. The other two families were rejoiced at this ſeparation and congratulated each other upon their departure. The wife was of a peeviſh temper, and her huſband was ſo jealous of her, that he would ſcarce leave her an inſtant; he watched even her ſlighteſt motions, and would infallibly have diſturbed the good underſtanding they were deſirous of preſerving among themſelves. A perfect union prevailed between the two families, that made the voyage with us, and were landed and ſettled by us on the Malouine Iſlands. One of them conſiſted of a man, his wife, two children, one a boy of three years old, the other a girl of one year, and two ſiſters of the wife, one twenty, and the other ſeventeen. The other family was compoſed of a man, his wife, a boy of four years old, and the wife's ſiſter,

ſiſter, about ſixteen. The wife was ready to lie in, when we left the iſland on our return to France.

In the morning of the 23d of September, the wind got to E. N. E. an eaſy gale. As it ſeemed ſteddy in that point, M. Duclos our Captain made a ſignal to bring in our long-boat, and yawl, which were on ſhore, the long boat to get water, and the yawl to fetch the ſailors, and the women that waſhed the linen. M. de Bougainville, M. de Belcourt, M. l'Huillier, and M. Donat were out in purſuit of game, near two leagues up the country, and propoſed to dine at the Caſtle of la Latte, where M. Mauclair and myſelf expected them till half paſt two. M. Duclos ſeeing that none of them came on board fired a gun, which haſtened the return of our ſportſmen; but as the time preſſed, and they had dined in the country, they would not make any ſtay at the caſtle of la Latte. We ſent the dinner on board again, where M. Mauclair, and I contented ourſelves with a ſingle glaſs till ſupper.

At three, ſignal was given to the Sphinx to weigh anchor. At ſix, our boats being embarked, we ſet ſail from Cape Frehel; and after ſeveral tacks to double the caſtle of la Latte, at nine we were North and South of the point of the Cape.

On Monday, the 25th, about four in the afternoon, we threw out a line with a double hook. The hook was ſcarcely in the water, before a fiſh, in ſhape and colour reſembling a mackrel, bit at it, and was taken. It weighed about thirty pounds, and had not two handfuls of entrails, liver, &c. All the reſt was ſolid fleſh, like that of the thunny, of which it had the colour and flavour. An excellent ſoup was made of it the next day. Several ſlices of it were brought up with different ſauces, and we found it very good: it is ſomewhat dry, but not ſo much as the bonito. It is called by the French, *Grand-Oreille*.

The hook, with which it was caught, was not baited with fleſh, fiſh, or any inſect. It is compoſéd of two ſtems of iron, about the thickneſs of the quill of a pen, faſtened together. They cover this double ſhank with tow, ſo as to give it the form of a ſpindle: the tow is covered with a piece of ſtrong white cloth and a plate of lead; to this they join two or four white feathers from the wing of a gooſe or fowl, placing them in ſuch a manner as to reſemble fins when extended. In this ſtate, the hook has nearly the appearance of a flying fiſh. The end of the ſhank is turned in a ring, through which they put a braſs wire of almoſt the ſame thickneſs, and about two feet and a half in length; the whole of this is thrown into the water, being faſtened to a cord about the thickneſs of one's little finger, and of the length of twelve fathom. One end of this cord is faſtened to the ſtern of the ſhip; the other, where the hook is, drags at a great diſtance in the track of the ſhip.

We continued our voyage for ſeveral days without any thing remarkable, wind varying, and weather generally ſtormy. We ſaw ſeveral ſhips at a diſtance, which we took to be on their return from the cod fiſhery on the banks of Newfoundland. One of them brought to, and ſpoke with us.

On the 2d of October, about nine in the morning, we deſcried a veſſel without maſts, and bore down upon her in order to give her what aſſiſtance we could. At ten we ſpoke with her. She proved to be a Dutch Merchantman of Amſterdam; ſhe was coming from Curaſol, and meeting with a guſt of wind at about a hundred leagues from Bermudas, they were obliged to cut away the mizzen and main maſt. We inquired if they were in want of any thing; they anſwered, that they had five French ladies on board whom they were carrying to France, but that they could not put their boat to ſea. We acquainted them, that we were juſt come from France, and ſhould not return thither for ſeveral months, for which reaſon we could not take charge of the ladies; but if they were in want of any thing elſe, they might come and fetch it. They again told us, that they could not put their boat to ſea. The ſea indeed ran high, and we not caring to expoſe ours to it, wiſhed them a better voyage and continued our courſe S. W. $\frac{1}{4}$ W.

The 5th, at break of day we diſcovered a ſail. We were in thoſe latitudes, where the Sallee Rovers ſometimes cruize; and we knew, they had a Frigate at ſea, called the Bird, of 36 guns and 300 men, which the Engliſh had ſold to the Salletines, and they had given the command of it to a renegade captain of Provence, a good ſeaman and of approved courage. They had alſo a ſloop of 12 guns and a hundred men. In conſequence of this, the commandant of our two frigates had iſſued out orders, that they might be able to act in concert, in caſe of an attack. The plan of the engagement was fixed up; the guns and ſmall arms were prepared; every man repaired to the poſt allotted him, and we bore down. It was ſettled, that if this was the Salletine frigate, the Sphinx ſhould hoiſt Engliſh colours, and ſeem to make all the ſail ſhe could to get under the fire of the frigate, to avoid falling into our hands. We in conſequence were to hoiſt French colours, and make a ſhew of purſuing the Sphinx, firing at her at the ſame time as if to bring her to. As ſoon as the Salletine frigate ſhould be got between the Sphinx and us, the Sphinx was to hoiſt French colours, and then make her a compliment of her whole broadſide, ſo that ſhe ſhould find herſelf between two fires. It was hoped, that by this manœuvre, we might make up for our want of numbers, and ſhatter her ſo by a vigorous attack, that ſhe ſhould be obliged to ſtrike.

Our men were brave fellows, and diſplayed at this time an air of gaiety and reſolution. They had indeed a great confidence in the ſkill and courage of our captains, and other officers, with whom they had made cruizes in the laſt war, and under whoſe command they had taken many prizes, and had even made themſelves maſters of ſome Engliſh ſhips at cloſe quarters.

As we neared the ſhip we had ſeen, we thought we could diſcover that ſhe was Engliſh built. But as we knew, the Engliſh had ſold ſeveral ſhips to the Salletines; and this, notwithſtanding we bore down upon her, hoiſted no colours, we took her for a Salletine ſcout. On this we fired a gun, and advanced upon her. Still ſhe hoiſted no colours. We now fired a loaded gun, and it is probable ſhe felt the wind of the ball. She then lay to for a moment, and afterwards ſtood for us, without hoiſting. When ſhe was got pretty near, ſhe hoiſted Engliſh colours, and paſſed ſo cloſe to us, that we diſcovered the captain to be the ſame Guernſey man, who ſerved as pilot to the Engliſh in the laſt war, when they made their deſcents at Cancale and St. Cas. The uſual queſtions were put to him in French, as, from what port, and whither he was bound, and what was the name of his ſhip. He made no anſwer. M. de Belcourt took the ſpeaking trumpet, and put the ſame queſtions to him in Engliſh, with all the embelliſhments of the emphatic ſea ſtyle, adding, that he deſerved to have had his ſhip ſunk for not hoiſting, after having been twice fired at. To this he replied in Engliſh, and alledged, that his colours had been entangled among the goods. It proved to be a merchant ſhip with two maſts, bound, as he told us, from Liſbon to St. Michael's, one of the Azores.

The 8th in the morning being calmed, we ſent out our cutter for M. de la Gyraudais, captain of the Sphinx. He came on board us at ſeven. M. de Bougainville, and M. du Clos our captain, had a conference with him. He received orders for his rendezvous in caſe of ſeparation, and exact drawings of the places we were to touch at, and of thoſe we expected to find in our courſe. M. de la Gyraudais returned to his own ſhip about nine.

The 9th and 10th, the calms continued with fogs, and ſome ſtorms of rain. The 11th the ſame. The currents here ſeem to run North; as may be conjectured from the difference we found between our reckonings and obſervation of yeſterday and to-day, in which time we had made ſeven leagues and a half of way. The 13th in the morning, the ſea being fallen after a ſtorm which roſe the evening before, we caught three fiſh called bonitos. There were not leſs than fifteen of them and two gold fiſh, playing about on the ſtarboard of our ſtern. We ſaw at the ſame time ſome other fiſhes which go under the name of pilots. One of theſe was taken in a net; the bonitos were caught with a line,

line, baited with the figure of a flying fiſh. Theſe weighed each of them about twenty pounds; the pilot was not more than eight inches long.

The 14th, being between 29° and 30° latitude, we expected to meet with the trade winds, of which we had hitherto had no ſigns. Some of our ſailors, who had the moſt experience, had aſſured us they were commonly found under this parallel. M. de Bougainville was ſo impatient for them, that he never ſtirred out of his cabin without going to examine the compaſs. He was obliged however to put up with ſuch winds as happened to blow.

At two in the afternoon, the Sphinx, which was to eaſtward of us, attracted our attention by hoiſting a white flag at the foremaſt head, which was the ſignal agreed upon in caſe of ſeeing land. We anſwered her with the ſame ſignal, and found it to be Palm Iſland, the fartheſt to the North Weſt of the Canary Iſlands. It bore E. S. E. of us by the compaſs, and appeared to us, at about 15 or 18 leagues diſtance, in the form it is repreſented in the plate.

We diſcovered another at the ſame time, more to the South Weſt, which exhibited nearly the figure B.

The ſight of theſe Iſlands was of uſe in correcting our reckonings and obſervations, and we found that we were about 20 leagues farther Weſt than we had reckoned.

The 16th, at three in the afternoon, we made a ſignal to the Sphinx, that we were going to make all our ſail; which we had not hitherto done ſince our departure, in order that ſhe might be able to keep up with us. The Sphinx was not near ſo faſt a ſailer as our ſhip, and had kept us back at leaſt a hundred leagues; but we did not chooſe to quit company ſooner, for fear of meeting with the Sallee Rovers, which would have required our mutual aſſiſtance to extricate us from them. At this time that we had got out of the latitudes, in which they cruize, we reſolved to ſtretch away for the place of rendezvous; that by arriving there as ſoon as poſſible, we might have all the refreſh-

ments,

ments, which the Sphinx might ſtand in need of, ready againſt her coming in, by which means our ſtay might be ſhortened.

As ſoon as the Sphinx had anſwered our ſignal, we ſet more ſails, the wind blowing freſh, and by ſix o'clock in the evening ſhe was at leaſt three leagues a ſtern of us; and before next morning we loſt ſight of her.

On the 18th and 19th, we ſaw a great number of flying fiſhes. They were purſued by thunnies and gold fiſh, which ſprang three or four feet out of the water to ſeize them. We threw out ſeveral hooks, but not one of them would bite.

During great part of both theſe days, our weather had been very ſtormy and the ſea ran high. On the morning of the 20th, a calm ſucceeded, with ſome rain at intervals.

Theſe calms and the winds which never blew freſh, and were continually changing, did not promiſe us a ſhort trip. We began all of us to be impatient at not meeting with the trade winds, which would have been ſo uſeful, and were ſo much the object of our wiſhes. M. de Bougainville particularly exclaimed againſt all former navigators, who have laid it down as a certainty, that thoſe winds never fail to blow in theſe latitudes. He told us, that, as we had experience of the contrary, he was reſolved on his return to Paris, to preſent a memorial to the Academy of Sciences, to prove the non-exiſtence of trade winds; at leaſt, the little dependence that navigators ought to have on what is related of their conſtant influence.

The 21ſt in the afternoon, we ſaw a great number of flying fiſhes, and of their enemies the bonitos, gold fiſh, and thunnies.

The morning of the 22d preſented us with about half a ſcore of flying fiſh, which attempting to fly over the frigate had fallen foul of the ſails, and dropped upon deck. They were dreſſed for dinner, and we found them extremely good and very delicate eating. I kept one in order to paint it from the life, the figure of it is to be found in the plate.

This fiſh in theſe latitudes is of a fine blue on the back, which fades or grows ſtronger inſenſibly towards the bottom of the belly,

belly, where the colour is a blue with a ſilver caſt. Its wings are fins of a greater length, which in general extend as far as the tail, but in ſome do not reach farther than to the middle of the body; though the fiſh is of the ſame ſhape, length and thickneſs. The one, whoſe figure is repreſented in the plate, was about ten inches from one extremity to the other.

On the 23d in the afternoon, ſome of the ſailors ſeeing a number of thunnies, got on the prow of the frigate with a harpoon, and caught one of them, which weighed 72 pounds. On a cloſe examination of it, I perceived ſome animals ſticking, and as it were glued upon its ears. See the figure of them in their natural ſize in the plate. The figure marked D is the upper part of the animal, which reſembled a compoſition of ſtrings of catgut almoſt tranſparent. Its eyes were two little black ſpots placed above the mouth B. They faſten themſelves on the thunny by means of two legs marked C, and two others conſiderably ſmaller marked D.

I took ſome ſea water and put it in a clean glaſs tumbler, that I might keep this animal alive and ſee its motions. I perceived in this water a black ſpot, which at firſt I took for a ſpeck of dirt; but when I attempted to take it out with the end of my finger, I obſerved the ſuppoſed atom to avoid my touch, and plunge under water. I attended to its motions, and found it to be a living creature of the ſtructure and ſize deſcribed in the plate. It was a ſpecies of cylinder formed by ten rings, ſo ſlight and tranſparent, that it was neceſſary to put the glaſs between the light and the eye of the obſerver in order to perceive it. It ſwam by means of two long fibres B B, and two others that are almoſt imperceptible C, which in gathering up and lengthening out again, gave the rings of the cylinder a motion perfectly correſponding to that of a quail-pipe, or a powder-machine uſed by hair-dreſſers. The body A, was of a violet colour towards C, and of a light brown towards B B.

We ſaw likewiſe a great quantity of flying fiſh, and we caught with a hook a bonito and a pilot, which I have painted from the life.

The

The naturaliſts pretend, on the authority no doubt of ſome ſeamen, that the pilot always goes before the ſhark, and that it is for this reaſon that fiſh has obtained the name of the pilot, as being director of the other's courſe. For my own part, I have ſometimes obſerved one or two pilots before or after each ſhark we caught; but we have often ſeen pilots without ſharks, as well as ſharks without pilots.

Father Feuillée, p. 173, confounds the pilot with the ſucking fiſh, and makes them both the ſame. " The ſharks, ſays he, are accompanied by little fiſhes, which keep continually with them, and chooſe rather to ſhare their fate than to abandon them; they ſwim always a head of them, at ſuch a diſtance that the ſharks cannot catch them, which has procured them the name of pilots. We did not catch a ſingle ſhark without finding ſome of theſe ſmall fiſhes ſticking to his back, by means of a yellowiſh, cartilaginous membrane of a circular form, which they have on the top of their heads: this membrane has an infinite number of ſmall holes filled with fibres, which, to all appearance, ſerve to draw from the ſkin of the ſhark ſome ſubſtance for their nouriſhment.

The ſame author allows the ſhark but three rows of teeth, one of which, he ſays, is compoſed of triangular teeth, and theſe are of a greater length than the others; I have counted ſeven rows of them in the mouths of all the ſharks we took, all of them moveable and triangular. Nor were the ſuckers of theſe ſucking fiſh of a circular, but of an elliptical form, ſuch as is deſcribed in the figure I have given of them in the ſequel.

The 24th the ſame winds continued, which we had had for ſome days. Theſe were in fact, the trade winds we had ſo long looked out for, under which name are comprehended all thoſe which blow from S. S. E. through the eaſterly point to N. N. E. incluſive. Theſe are the moſt favourable winds that can blow, for ſhips bound from Europe to South America, the windward and leeward Iſlands, and the Gulph of Mexico.

About eight in the morning on the 25th, we had ſight of land on our ſtarboard ſide. At noon, we judged it to be the

7 Iſland

Iſland of Bonaviſta, one of the Cape de Verd Iſlands, ſituated North Eaſt of St. Jago, the largeſt and moſt populous of thoſe iſles. It bore North Weſt of us, about nine leagues: the figure of it, in the moſt extenſive view we had, appeared according to the repreſentation in the plate.

This iſland, like the reſt, abounds in wild horſes, goats, and ſeveral other animals, notwithſtanding the ſoil is rocky and barren. It is ſeen at a great diſtance by means of its white cliffs, from which circumſtance it derives its name.

We now found that we were near twenty leagues farther eaſtward than our reckoning.

The wind blowing freſh from N. E. to N. N. E. accompanied with fine weather, afforded us a view of another of the Cape de Verd Iſlands, about four o'clock in the afternoon, which goes by the name of Mayo's Iſland. The ſoil here likewiſe is rocky and barren. There are nevertheleſs a great number of bulls, cows, goats and aſſes. A conſiderable quantity of ſalt alſo is made here. The air is hot and unhealthy. The moſt ſouthern point of the iſland bore S. W. $\frac{1}{4}$ W. and the moſt northern W. $\frac{1}{4}$ S. W. of us, and the whole appeared as exhibited in the plate.

The 27th, after having had ſome lightning in the night, and in the morning a cloudy ſky, with a high ſea, and a ſquall of wind at half paſt ten, ſucceeded by a ſtorm at E. S. E. which was of ſhort duration, the wind came about to the uſual points with an eaſy gale; and about three in the afternoon, we caught a bonito, which weighed forty pounds.

The 28th and 29th, proved very ſtormy, but notwithſtanding this we were not driven out of our courſe.

On Sunday the 30th, in the morning, the ſky cleared up and the wind came fair again.

At eight o'clock, Peter Lainez of St. Malo, a cabbin boy, about twelve years of age, going into the forecaſtle fell overboard, without any one's knowing how the accident happened. The ſecond mate, who was going a ſtern, ſeeing him float along the ſtarboard ſide, cried out immediately, that there was one of

F the

the crew overboard. We ran at this time four knots an hour, with a quarter wind. They threw out a plank directly from the ftern gallery, and whatever elfe was at hand either of board or any other buoyant materials, in hopes the poor fellow might be able to reach fome one of them, and keep himfelf by that means above water, till the boat could be put out to take him up. The whole crew was in motion, and every poffible means were ufed to ftop the fhip. Many ran up the main maft, others got on the quarter deck, all intent on looking for, and difcovering the cabbin boy. After this, the boat was put to fea though it was then very rough; it was manned with fix ftout failors, under the command of the mate, who went in fearch of the cabbin boy to the right and left, wherever they thought they had a chance of finding him, to the diftance of half a league from the fhip, but without fuccefs. When they had been out about three quarters of an hour, a fignal was made for the return of the boat, which was effected with much difficulty. We reembarked her, and continued our route.

The names of the crew were then called over, in order to find out who was the perfon miffing; for we did not yet know that it was the cabbin boy I have juft now mentioned. He was the only one, that did not appear. They looked in his hammock and fearched the whole fhip over for him, and not finding him any where, it was eafy to conclude, that this Peter Lainez was the hand we had loft.

At four in the afternoon, after vefpers, the cloaths of the deceafed cabbin boy, an inventory of which had been taken in the morning, were fold by auction. Our commandant, M. de Bougainville, bought almoft every thing, and diftributed them gratis among the cabbin boys, who were leaft in a condition to procure any for themfelves. The fale amounted to fifty crowns.

The 31ft, the weather was ftormy at intervals, each ftorm being fucceeded by an almoft dead calm. During thefe calms we caught, in lefs than two hours, two fharks, which weighed about a hundred pounds each. They had both of them fifhes fticking ſticking to their bodies near the head. Theſe fiſh are called ſucking fiſh. I painted one after the life, in two figures; the firſt ſhews the ſide of the ſucker, which is upon the head; the other figure repreſents the belly of the fiſh. It was ſeven inches in length.

A few hours before, ſome hundred porpoiſes, whoſe figure may be ſeen in the plate, made their appearance within piſtol ſhot, and ſeemed as if they had come on purpoſe to amuſe us. They ſprang out of the water in an extraordinary manner. Several of them in cutting their capers, leaped at leaſt three or four feet high, and turned round not leſs than three times in the air, as if they had been on a ſpit. One may judge from hence of the ſtrength of this fiſh.

On the 2d of November, at three in the afternoon, a ſtorm roſe at South Eaſt attended with a heavy rain. During this ſtorm one of the ſailors brought me a flying fiſh, eight inches and a half long, which had juſt fallen on the forecaſtle. We had ſeen, before the ſtorm came on, ſhoals of thunnies and bonitos. They leaped out of the water, and made the ſea foam, as if they were fighting with each other.

On the 3d, a ſhark of a middling ſize, and about a hundred and fifty pound weight, came a ſtern of us. He bit at the bait, as ſoon as it was offered to him. When he was raiſed out of the water, he gave a ſudden jirk, by which he diſengaged himſelf from the hook, leaving part of his jaw behind him. Not diſmayed or diſheartened by this loſs, the ſhark perceiving the ſame piece of bacon, which had been made uſe of as a bait for him the firſt time, thrown out again, returned to it with the ſame greedineſs, and ſwallowed at once not only the bacon, but the piece of his jaw, without however being caught by the hook. Another piece of bacon was immediately put on: the ſhark was without doubt very hungry, for he came again to ſeize that. But as at this time there was a dead calm, and beſides, this fiſh is neither wholeſome nor palatable food, inſtead of endeavouring to take him, we amuſed ourſelves near an hour with

 only

only letting him ſmell the bait. When he attempted to ſwallow it, we drew it ſuddenly out of the water, and this was repeated at leaſt a dozen times without producing the effect, which is ſaid to be ſo common upon theſe occaſions, of making the ſhark ſpring out of the water in order to ſeize it.

Another thing I muſt obſerve, is, that I did not ſee him turn upon his back to ſwallow the bait, but only a very little on one ſide. M. de Bougainville, while we were amuſing ourſelves in this manner, fired at him twice with muſquet ball, but whether he miſſed him, though almoſt at the muzzle of his piece, or whether the ſkin was too tough for the ball to penetrate, the ſhark was not in the leaſt diſturbed in his motions by it; he kept ſwimming round and round the ſtern, and at laſt ſwallowed this ſecond bait without being hooked. A ſquall of wind riſing about this time, we left the ſhark to employ himſelf elſewhere.

The 4th and 5th, we had ſtorms and calms at intervals. The 6th, about ten at night, we had a ſquall of wind, which cleared the ſky. At this time we ſaw ſome ſtars; a ſight we had not had for near a week, the ſky having been always gloomy and covered

The morning of the 7th, the ſun roſe fine, but with ſeveral clouds ſcattered round it. Before it appeared, the rays darting upon theſe clouds exhibited one of the moſt beautiful ſights in the world for variety and brightneſs of colours. I was mortified more than can be imagined, not having it in my power to paint ſuch a day-break, which would have made a moſt brilliant picture. I have only been able to preſerve a very imperfect ſketch of a ſetting ſun, which we all of us admired for near half an hour. But it is not poſſible with water colours to execute a picture, upon which any exact idea of it might be formed. Theſe colours are too faint to expreſs the brilliancy and luſtre, with which the borders of the clouds were illuminated by the rays of the ſun. Oil colours would without doubt be leſs defective in the repreſentation; but I had not any with me.

Beſides,

Befides, it would require a fkilful painter to execute fuch a picture properly; and I have not that qualification.

The weather continuing fine and very hot, we had all the hammocks between decks taken down, in order to dry the cloaths of the crew, which had been all wetted in the rainy days. This dampnefs of cloaths is a much more immediate caufe of the fcurvy, and many other diforders, than the falt provifions which are ufed at fea. A captain cannot pay too great an attention to the preferving of cleanlinefs among his crew, and to the airing of the hammocks, cotts, &c. if he would prevent diforders. Our captain affured me of the truth of this obfervation, from his own experience in the different voyages he has made to China, India, Peru, and Canada. He told me, he had always paid ftrict attention to this article, to which as well as to the choice of proper food, he attributed the general good health his crews had enjoyed during thofe voyages.

In the afternoon, we faw a large bird called by fome *Goellan*, or *Gull*, and by others *Caignard*. At night a fingle fwallow came and perched on the main maft yard, and the next morning continued flying round the fhip.

During the night feveral flying fifh dropped upon our deck. They were all of that fpecies, which have the fins, that ferve them for wings, reaching to their tail.

At five in the morning of the 9th, a bird pretty nearly of the fize of a pigeon, but fomething longer, coming to perch on the foremaft yard, one of the failors caught him in his hand. This bird, which I have painted, and whofe figure in half the natural fize may be feen in the plate, is of a light brown inclining to red, almoft the colour of a nut. The largeft feathers of the wing and tail, are of a darker brown, or rather blackifh. The bill is black, ftrait, and fmall, pierced through in the middle, with a fmall protuberance below, about the length of the bird's head. The upper part of the head near the bill is white; it then becomes of a pearl colour, growing deeper towards the neck, which is pretty long in proportion to its thicknefs. The feet are

of

of a dark grey, webbed like thoſe of water fowl. After having made uſe of this bird in the manner I ſhall mention hereafter, M. de Bougainville gave him to me to paint. I put him in a ſmall preſs in my cabbin, where I found him the next day very lively, and ſo little alarmed at having been taken, that when I ſet him on my table, he put himſelf in the attitude, in which I have drawn him. I gave him ſome food, and he eat of it, ſtill keeping in the ſame poſture, and continued ſo for three days, by which means I had full time to paint him to the life. Some of our ſeamen ſaid, it was a ſpecies of the *booby* bird, becauſe it ſuffered itſelf to be caught in the hand, and grew tame, as ſoon as it was taken: but he had not however the crow bill, which belongs to the booby, and has procured it the name of the duck with the narrow bill. Our ſeamen gave the ſame name to another bird alſo, very much reſembling this, except that it has a crooked bill, like that of a parrot.

About ten o'clock in the forenoon the ſea appearing of a light green caſt, inſtead of its uſual blue, and the colour continuing the ſame at ſix in the evening, we ſuſpected that this appearance was occaſioned by our being in the neighbourhood of ſome land, or ſhoal. We therefore took the precaution of ſounding; but though we ſounded with a hundred and twenty fathom of line, we found no bottom. Thus we were freed from the apprehenſions we had entertained, and which aroſe from an error of the charts; almoſt all of them placing Brazil near fifty leagues farther Weſt, than it is found to be by the obſervations of our ſeamen. We reſolved however to ſound a ſecond time, if the ſea had continued of the ſame colour; but as it appeared the next morning of its uſual blue caſt, we continued our courſe without taking the trouble of ſounding.

Our mates, boatſwain, and thoſe of the crew, who in former voyages had paſſed the line, had for the laſt week been making preparations for *the ceremony of Baptiſm*, which is performed on the part and in the name of the *Bon-homme la ligne*, to all thoſe, who

who have never before paſſed the line, without diſtinction of rank, or quality, or exception of perſon.

About ſeven o'clock, as we were at ſupper, we heard the ſmacking of a whip, which announced to us the arrival of a courier from the *Bon-homme la ligne*, according to cuſtom, the evening before the ceremony I juſt now mentioned is to be performed. This was the cockſwain properly equipped for a courier. He knocked at the cabbin door. We called out, who is there? A meſſenger, ſays he, from the *Bon-homme la ligne*, lord and governor of theſe latitudes. Let him in, ſays M. de Bougainville. The door was opened, the meſſenger alighted, and came in, leaving his equipage at the door. This equipage was formed by two ſailors tied back to back, and going upon all fours. One of them had on his head a ſwab, to repreſent the tail of the beaſt, the other had one for the mane, and a maſk of paſteboard in the ſhape of a horſe's head. The furniture conſiſted of the quarter cloths belonging to one of the boats; that is to ſay, of a carpet, or large piece of blue cloth, adorned with flowers de luce made of yellow ſtuff.

The meſſenger being introduced addreſſed our Commandant in the following terms: " *the Bon-homme la ligne, lord governor of theſe latitudes,* underſtanding, that the brave Chevalier de Bougainville, commander of the Eagle frigate, is arrived in his dominions, has ordered me to come and compliment him on his part, to let him know with how much joy he hath received the news of his arrival, to bring the beſt wiſhes for his health, and to deliver him a letter, in which my maſter hath expreſſed his own ſentiments.

M. de Bougainville read the letter, which was conceived in the following terms; *Brave Chevalier, your illuſtrious actions have rendered the French name highly celebrated in Canada: your renown has reached the latitudes over which I reign, on the wings of fame, and the hearts of my ſubjects are ſo filled with veneration for you, that the gold fiſh and bonitos, the thunnies and porpoiſes as ſoon as they deſcried the frigate Eagle, which you command, came in ſhoals to me yeſterday*

to

to announce your arrival. The joy, with which your presence had animated their hearts, they expressed by repeated bounds and leaps, which they continued for a long time as they passed by your ship. I send this ambassador to notify to you my own particular share in the general joy, at the same time that he delivers this letter into your hands, and I hope to-morrow to acquaint you in person, how much I am delighted with the visit you pay me.

Signed BON-HOMME LA LIGNE."

Given at the 54th minute of the first degree of latitude, and in 29 degrees three minutes longitude, of my northern dominions, the 9th day of November in the year of my reign, 7763.

M. de Bougainville, when he had read the letter, told the envoy, that he expected to have the honour of presenting himself before the *Bon-homme* the next day, and of giving an answer to his letter in person. Let the courier drink, added he, and take care of his horse: it must be a fine one; lead it in, I have a great curiosity to see it. The horse was introduced curvetting, tossing his head, pawing, and neighing. As it was possible he might be tired with his journey, and might be thirsty, a glass of wine was offered him, which he drank. The courier informed us, that his horse had two heads, one at his stem and the other at his stern, upon which the head at his stern also was treated with a glass of wine.

The courier, before he retired, presented to the commandant a bird on the part of the *Bon-homme la ligne*, the illustrious president of these latitudes having requested his acceptance of it as a token of his goodwill and affection. This was the bird which they had just before caught in the hand, and which I have mentioned above. But as we knew nothing of this at the time, we were not a little surprised at the present. We took it at first for an artificial bird, till by pecking with his beak he convinced us, that he was not only a real bird, but also in full vigour. On examination, we found it to be a fresh water bird, which served only to increase our surprize.

After ſupper, we aſſembled on the quarter-deck, and danced minuets, country dances, &c. to the tabor, and after that to two violins till near ten o'clock; when we retired to our cabbins.

Thurſday, the 10th of November, at five in the morning we paſſed the line, at 29 degrees 3 minutes longitude according to our reckoning. At ten o'clock we ſaw a bird called the *Frigate*. This bird is frequently found at four hundred leagues diſtance from land, though it is ſaid not to be able to ſupport itſelf on the water without periſhing, which is the caſe with birds, that are not uſed to live in that element. Its legs are ſhort, thick, and gathered up cloſe to the body. Its feet are not webbed, but furniſhed with ſtrong pointed claws. Some of them meaſure nine feet from the tip of one wing to that of the other. By the extent of its wings when they are ſpread, this bird eaſily ſupports itſelf in the air, the motion it gives them being almoſt imperceptible. Sometimes it riſes to ſo great a height, that the ſtrongeſt eye loſes ſight of it. When it comes near any ſhips, it flies round the vanes of the maſt head, going and returning very frequently, but never perching on any part. The ſize of it is nearly the ſame as that of a fowl. Its look is ſteady and piercing. It darts upon its prey with an incredible ſwiftneſs and ſeizes it both with its talons and bill, the upper part of which is unciform. The males have a red granulated membrane deſcending from their bill as far as the middle of their neck. The feathers on the belly are of a light grey, which at a diſtance make it appear white. Thoſe on the back and wings are brown. This bird ſaw ſome flying fiſh, which he caught very artfully, by ſkimming along the ſurface of the ſea, while they were flying to avoid becoming a prey to the bonitos, and other fiſhes, which are enemies to them. It is ſaid, that he purſues the gull likewiſe and other ſea birds, to make them diſgorge the fiſh they have ſwallowed that he may ſeize upon them himſelf.

I do not well know for what reaſon this bird is called the frigate, unleſs it be by way of compariſon between the ſwiftneſs of his flight, and the lightneſs of thoſe ſhips, which go under that name, and are uſually better ſailers than any others.

Not having been able to get a nearer view of this bird than from the top of the maſts, I cannot pretend to give a deſcription of it otherwiſe than from thoſe perſons, who have ſeen and handled them. Father Labat (Nouveaux Voyages, tom. 6. p. 395) in addition to what I have already obſerved, ſays, that this bird has large black eyes; that he ſeldom alights on the ground, and generally keeps himſelf perched, becauſe the ſize of his wings, and the ſpace which is neceſſary in order to put them in motion, would render it very difficult for him to riſe from the ground. The feathers on his back and wings according to the ſame author are black, thick and ſtrong; and thoſe which cover the ſtomach and thighs, are more delicate and not ſo black: perhaps that which I ſaw was the female, or at leaſt a young one. I killed ſome of them (continues the father) in the iſland where we were, for their greaſe. It is ſaid to be an admirable ſpecific in the ſciatica, and in a numbneſs of the limbs, and other accidents ariſing from a want of circulation. The greaſe is to be heated, and while it is on the fire, the parts affected are to be well rubbed and chafed in order to open the pores; and ſome good brandy, or ſpirits of wine are to be mixed with the fat immediately before it is applied. A piece of blotting paper, ſteeped in this mixture, may be laid on the part, with compreſſes and a bandage to keep it in its place.

We now come to the baptiſm of the line.

About two o'clock in the afternoon, they began by placing a bathing tub full of ſea water and two buckets on the quarter-deck: athwart, from the ſtarboard to the larboard ſide, they ſtretched a rope, which they called *the line*, the ſame they uſed for ſounding; then the drum was beaten for every body to aſſemble. The weather proved very ſeaſonable for the ceremony, for it was extremely hot. Near the gangway, which leads to the ſtate room, was placed a bench covered with the quarter cloths, which had ſerved the evening before to capariſon the couriei's horſe; and this was to be the ſeat or throne for the lord governor of the line, his chancellor, and the vicar, who was to adminiſter the baptiſm.

When every body was aſſembled, a voice conveyed through a ſpeaking trumpet called out from the main maſt top; *what is the name of this ſhip I ſee below within my dominions? The Eagle,* anſwered the captain.—*Who commands her?*—M. le Chevalier *de Bougainville.—I am very glad of it; it will give me pleaſure to admit him into my ſociety, according to the eſtabliſhed forms and ceremonies. I received an account of him yeſterday, and as a teſtimony of my ſatisfaction, am coming down into his ſhip with all my court.— A la bonne heure,* replied M. de Bougainville, a ſea phraſe to expreſs, that one underſtands what has been ſaid, and that one approves of and conſents to it.

Upon this a ſailor, who had no other covering than a pair of tarred breeches, and on his ſhoulders the ſkin of a ſheep with the wool on, ſtained with red and yellow in large blotches, with a cap on his head made likewiſe of the ſkin of a ſheep painted, with a pair of bull's horns faſtened on the top, and ſeveral pieces of wood blacked, and feathers of turkies and fowls upon it; his breaſt, arms, legs and face, being ſtained in the ſame manner with red and yellow colours, diluted in oil, and large black whiſkers; this ſailor, I ſay, thus accoutred came down from the main maſt top by the ſhrouds on the larboard ſide, with an iron chain round his middle by way of a girdle; in one hand he held the end of this chain, and in the other a pot-hook.

Six cabbin boys marched before him naked, painted from head to foot with red and yellow, ſome of them in blotches, others in croſs bands after the manner of the ſavages.

As ſoon as they came on the quarter-deck, the ſailor drew them up in order, placed their thumbs on the rope, and made them dance for a quarter of an hour to the tabor. After this they approached the bathing tub, and the ſailor threw ſeveral buckets of water over them.

This

This ceremony being finiſhed, the deſcent of the lord governor of the line was announced by the throwing of white kidney beans, for ſugar plumbs, from the main maſt top on the quarter-deck. The Bon-homme la ligne, preceded by his whole court, took the ſame route as the ſailor and the cabbin boys; he deſcended ſlowly and majeſtically. His court was compoſed of the ſecond mate, the boatſwain, the pilot, and the gunner. The firſt mate repreſented the Bon-homme la ligne. He was covered with white ſheep ſkins ſewed together ſo as to make a garment of one piece. His cap, which was compoſed of the ſame materials came down over his eyes. A quantity of tow mixed with wool ſerved him for a peruke and a beard. He had a falſe noſe made of painted wood. Inſtead of a ribband, he wore acroſs his ſhoulders a ſtring of trucks of the parrels, as large as gooſe eggs.

His attendants were dreſſed up much in the ſame manner, except that ſome of them had their arms or their legs naked, and painted red and yellow, as likewiſe their faces ornamented with large black whiſkers, and long wooden noſes. One carried a mace, or club ſuch as the ſavages uſe, another a bow, a third an ax, and a fourth a calumet. Near the lord governor was his chancellor bearing the ſcepter, which was a ſort of mop, ſuch as is uſed in ſpunging a cannon, after it has been fired. The cockſwain dreſſed like a woman, and painted with coarſe red paint mixed up in oil, ſtood cloſe to the Bon-homme, who called him his daughter. As to the vicar, he was cloathed in a ſort of linen robe, covered with pitch and tar; a cord about the thickneſs of one's thumb ſerved him for a ſaſh. He wore a ſquare cap of paſteboard blacked over, a maſk of the ſame, and a linen gown painted red, and carried a book in his hand. One cabbin boy had a ſquare cap painted red and black, another held a wooden cenſer, hanging by pack threads platted in the ſhape of a chain, and in the other hand a chafing diſh with fire to heat the perfumes, which were made of pitch and tar. A third

cabbin

cabbin boy carried a bow and an arrow; and a fourth a bafon and watering pot full of fea water for the baptifm.

The whole proceffion being come down upon the deck, and the crew affembled there, the lord governor defired a conference with the commandant, who immediately advanced to receive him. *You are welcome hither, M. le Chevalier; I am happy to fee you,* faid the Bon homme la ligne: *excufe me if I do not make you a long compliment; my lungs are fo feeble, I can fcarcely fpeak. You muft not be furprifed at this; for I am 7763 years old: it is even with difficulty that I can write. I have therefore ordered my fecretary to do it for me; and here is a letter, which will acquaint you with every thing I had to fay to you, as well as my chancellor. I am come down from my palace on purpofe to admit you into my fociety. I hope you will make no fcruple of fubmitting to the ceremony of being baptized agreeable to the cuftom on this occafion.* M. de Bougainville received the letter, read it, and replied *à la bonne heure.* After this he faluted the daughter of the Bon-homme, and after congratulating him on his having fo handfome a daughter, drew near the line, or rope, which was ftretched acrofs. The officers of the Bon-homme accompanied him to it, and the lord governor feated himfelf on his throne with his daughter and his chancellor.

The officers tied M. de Bougainville's left thumb on the line with a red ribband. The reft of us gathered round, viz. Meff. de Nerville, de Belcourt, l'Huillier and myfelf, and they tied our left thumbs with the fame ribband.

The vicar with a folemn air, and with his book in his hand, approached M. de Bougainville. At the left hand of the vicar was the fcepter-bearer of the lord governor; and at his left hand two cabbin boys dreffed like favages; one of whom carried a plate covered with a napkin folded, to receive the tribute, which is called *ranfom,* becaufe they content themfelves with pouring a fmall quantity of fea water on the heads of thofe, who ranfom themfelves, inftead of plunging them in the fea, as is done in the punifhment of ducking: the other held a bow in one hand and

and a cenſer in the other. The cenſer was a piece of wood, hollowed in the ſhape of a porringer, with three handles, and ſuſpended by three pieces of cord. The cuſtom of dipping in the ſea in performing this ceremony of baptiſm is aboliſhed: it having been conſidered that that practice might be attended with much danger on account of the ſharks, which are apt to lurk near the ſhips, and carry away a thigh at leaſt from any unfortunate perſon, whom they happen to ſeize. In lieu of this, they have ſubſtituted the baptiſm of the bath, or bathing tub, on the edge of which they cauſe the perſon to ſit, who has not ranſomed himſelf, or whom they have a mind to plague, as will be ſeen in the progreſs of this account.

Things being thus ſettled, the vicar addreſſed himſelf to M. de Bougainville in the following manner: "In order to be admitted into the noble and puiſſant ſociety of the lord governor of the line, it is neceſſary to enter into certain preliminary engagements, which you will promiſe to obſerve. Theſe engagements have nothing for their object but what is entirely reaſonable." "A la bonne heure," replied M. de Bougainville. "Do you then promiſe," purſued the vicar, "to be a good citizen, and to that end to labour at the work of population, and not to ſuffer young women to languiſh away their time, whenever a favourable opportunity ſhall offer itſelf?—I do promiſe—Do you promiſe never to lye with a ſailor's wife?—I do promiſe.—Do you promiſe to cauſe the ſame engagements to be taken, and the ſame, or ſimilar ceremonies to be obſerved by all thoſe who have not paſſed the line, when they happen to be with you?—I do promiſe.—Put your hand then upon this holy book in token of your obligation." M. de Bougainville laid his hand on a cut, which repreſents a genius or angel and a young girl tenderly embracing each other. It is the cut at the 47th page of a book intitled, *Sentimens d'un Chretien, touché de l'amour de Dieu.* At the bottom of the cut is this ſentence: *quis mihi det te fratrem meum ſugentem ubera matris meæ & inveniam te foris & deoſculer te. Cant.* 8. The vicar went to the lord governor of the line, and reported

reported to him that M. de Bougainville had taken the engagements: to which the Bon-homme anſwered: *dignus eſt intrare in noſtro docto corpore: admittatur.* The vicar then returned to M. de Bougainville and ſaid; the lord governor of the line is pleaſed to admit you into the ſociety of which he is the head, and has ordered me to receive you therein by adminiſtration of his baptiſm. What is your name? Louis, ſaid M. de Bougainville. Very well; *ego, nomine reverendiſſimi domini domini & sereniſſimi præſidentis æquatoris te, Ludovice, admitto in ſocietate ejus.* In pronouncing theſe words, he ſprinkled over his head ſome drops of ſea water. Then they untied M. de Bougainville's thumb, who put ſome money in the plate under the napkin, and the vicar threw incenſe on him. After this the vicar proceeded to M. de Nerville, to whom he propoſed the ſame queſtions, and after him to the other paſſengers and officers with all the ſame ceremonies.

It was now come to the turn of a midſhipman, who was a ſad dog, and hated by almoſt every body. The vicar told him, that the lord governor had given orders for his being admitted with all the ceremonies in form. In conſequence of theſe orders, he threw one end of his robe over the fellow's head, muttered a few words, and afterwards gave him the robe, which had been freſh painted in oil, to kiſs. He then took ſome blacking, mixed with oil, in a ſmall pot born by one of the cabbin boys, and ſmeared his forehead and cheeks with it. This being performed they untied his thumb from the line, and conducted him to the bath, on the ſides of which were two notches large enough to receive a ſtick, that was laid acroſs, and was to ſerve as a ſeat for him. He had no ſooner ſat down, than they ſuddenly withdrew the ſtick from under him, and he fell with his poſteriors into the water, the tub being about half full, to which there was a cord likewiſe adjuſted in ſuch a manner, that by pulling one end of it, at the inſtant the catechumen tumbles in, it faſtens round his middle, and keeps him under, without his being able to diſengage himſelf, till the by-ſtanders are pleaſed

to

to give him his liberty. As ſoon as the midſhipman was nooſed, they ſmeared his head and face all over with black and red: after that they threw at leaſt five or ſix buckets of water over his head, and then ſuffered him to go about his buſineſs.

After this they came to the two Acadian girls. The vicar aſked them, if they were virgins? they ſaid, Yes. Do you promiſe then, ſaid he, to preſerve your marriage vow inviolable, in caſe you ſhall have a ſailor for your huſband? The promiſe being made, he juſt marked their foreheads, noſes, cheeks, and chins with black, in the ſlighteſt manner poſſible, and then poured ſome water over their heads, after which they retired. The ſiſter of one of theſe had hid herſelf in order to avoid this wetting. She was found however, and they were going to oblige her to ſubmit to the ceremony; but the vicar being apprized, that there were reaſons, why ſhe ſhould not be expoſed to that part of it, which was to be performed with the water, told her, that he would content himſelf with making ſome patches upon her face. She ſubmitted to this, and he kept his word. The two married women were not baptized, becauſe their children, who were too young to be left by themſelves, were ſo affrighted at the groteſque figures of the attendants on the Bon-homme la ligne, that they could not be pacified or brought out from the corners where they had hid themſelves.

Several others were afterwards baptized and bedaubed with black and red, but none of them were ſeated on the tub; becauſe when the others had begun to throw ſome buckets of water over them, they, to be even with them, returned the compliment. Thoſe who had been wetted, choſe to wet others: the ſtruggle was who ſhould throw moſt water, ſo that all thoſe who remained on the deck were as wet, as if they had been dipped in the ſea. But they were not ſatisfied with ſluicing one another; thoſe who had had their faces blacked rubbed them againſt others who had not undergone that ceremony, and by this means there was ſcarce a man in the whole ſhip's company who eſcaped a daubing; and they did not give over the ſport, till

they

they were all tired. This proved an unlucky circumſtance for the Bon-homme and his attendants, who loſt part of the tribute they would have received from thoſe, who were not baptized with the ordinary ceremonies. The reſt of the day was paſſed in dancing, and other kind of amuſements.

This farce is performed in every European ſhip on paſſing the line. But there is no preciſe uniformity obſerved in the ceremonies uſed upon this occaſion. Each nation has invented ſuch as are moſt conformable to its genius and character; and every ſhip is regulated according to the degrees of humour in thoſe who happen to preſide. Sometimes the perſon, whoſe office it is to adminiſter the baptiſm, gives each perſon a name taken from ſome bay, ſome cape, or ſome remarkable promontory on an iſland or coaſt; taking care at the ſame time to apply them in ſuch a manner as to expreſs the character, temper, figure or diſpoſition of the perſon ſo named. The ceremony is in general called the *baptiſm*, or the *ranſom*: the baptiſm, becauſe of the water thrown over thoſe, who are then paſſing the line for the firſt time: the ranſom, on account of the tribute, which is paid by thoſe perſons who are not willing to be wetted. The tribute is uſually whatever the perſon, who pays it, thinks fit to give. Sometimes it is impoſed by the actors themſelves: however they always take care to make their levy proportionable to the circumſtances of the perſons, from whom the tribute is exacted. Thus it is not always required in money, but ſometimes in wine, or brandy, or hams, or ſuch like; as when the captain of the veſſel, who is not exempt any more than his paſſengers, paſſes the line for the firſt time.

When the ſhip is not to paſs the line, but only the tropic, thoſe of the crew, who have already paſſed it, not being willing to loſe the tribute, which they look upon as their due, have taken it into their heads to call the tropic, the *eldeſt ſon of the Bon-homme la ligne, preſumptive heir of his poſſeſſions.* Upon the ſtrength of this they play the ſame farce at paſſing the tropic, that others do on paſſing the equator. They have even thought fit to perform this ceremony, when a ſhip for the firſt time doubles Cape St. Vincent to paſs the Straits of Gibraltar. The ſhips which are employed on the cod fiſhery obſerve the ſame practice, when they come within ſight of the great bank of Newfoundland.

From this time for ſeveral days we had nothing remarkable.

On the 14th, by our reckoning and obſervation we ſuſpected, that the tides and currents ſet ſouthward, agreeable to the remark made by the author of admiral Anſon's voyage.

At eight o'clock in the evening of this day, a bird ſimilar to that of which I have given a figure in the preceding plate, ſuffered himſelf to be caught in the hand over my cabbin. We ſhut him up in a hen-coop.

The next morning one of our boatſwains having taken him out of his place of confinement to put him upon his hand, the bird took wing and flew away. A ſhort time after we diſcovered a *frigate*: this bird kept wheeling round our weather flag, and ſeemed to peck at it more than once. We made the ſame obſervation on the currents this day as the day before.

We found this climate much the ſame as that of France in the month of May, the mornings and evenings being rather cold, though we were under the torrid zone; nor did we experience any of that burning heat, which is complained of in the relations of ſo many perſons, who have ſailed through theſe parts. It is true that ſince we had paſſed the line, we had always had ſome little wind at leaſt, had never been ſurpriſed by calms, and had been ſecured by the clouds from the rays of the ſun. Whether it were owing to our cleanlineſs, or to our frigate being new we were not troubled with thoſe inſects, which are mentioned in the ſame accounts; nor had we to this time one perſon ſick on board. In order to contribute to the preſervation of health, every evening after ſupper, the ſailors were ſet to dancing on the ſtern-caſtle. And indeed they were ſo diſpoſed to jollity, that they would play at hot cockles, hunt the ſlipper, or any other game, that promoted exerciſe and encouraged

mirth.

mirth. Some of them, who were naturally of a comic turn, would dreſs themſelves up in maſquerade, aſſuming very groteſque figures, and would paſs in proceſſion, or make their appearance in groups on the ſtern-caſtle, where they would dance minuets, cotillons, allemandes, country dances, and hornpipes. Moſt of them had learned theſe dances, while they were priſoners of war in the ports of Great Britain. The greateſt part of them had made their eſcape from thence at the riſque of their lives, in neutral veſſels, fiſhing boats, and even ſmall boats, which they found means to carry off. Several of them have aſſured me, that the Engliſh connived at their eſcape, and would even bargain with the neutral veſſels for their paſſage, or ſell them boats: that ſome lent them cloaths to diſguiſe them, others advanced them money, others again gave them money out of charity, and others furniſhed them with letters of recommendation to their friends in London, or in ſuch ports, where they thought the priſoners might embark with the leaſt danger. They even went farther; and in order to give them the means of living comfortably in the priſons, where they were confined, made them preſents, and paid them very liberally for little toys, which ſome of them employed themſelves in making, even to the buying of them little images of the Virgin Mary, of Saints, &c. made out of wood, and as ill ſhaped as may be imagined, where the artiſts had no other tools but their knives, and had never learned the trade. One of our crew, who had amuſed himſelf in this way, has told me more than once, that they would give him to the value of half a crown for one of his figures, with this caution only, not to boaſt of it among the Engliſh. A fine leſſon of humanity and charity!

Mirth and cleanlineſs are two points, to the promotion of which ſea captains ought to pay great attention. They contribute in no ſmall degree to prevent all thoſe diſorders to which ſeamen are uſually ſubject. For the ſame reaſon they ought always to mix a little vinegar with their daily allowance of water, which they put in a caſk, called *charnier*. What was ufed for the chamber, or ferved up at the officers mefs was put into great earthen veffels, which were filled to the height of half a foot or more with fmall pebbles. After the water has been drawn off from the cafks into thefe large veffels, called *jarrs*, which are expofed to the open air on the ftern-caftle or thereabouts, it is left there to purify for three or four days before it is drunk. It is imagined that the pebbles ferve to clear it from flime.

I muft not omit to obferve here, that the water we had taken on board at St. Malo, had not fuffered the leaft change, as it ufually happens between the tropics. Our bifcuit was equally well preferved. There were only fome pickled cabbage, and fome fmall cafks of veal, which were rather fpoiled: and that probably was more owing to a fault in the feafoning of them, than to the heat of the climate we were in.

On the 20th of November at eight in the morning we took a porpoife of about a hundred weight. I painted him from the life, but without preferving any proportion to his bulk. For the figure of him fee the plate annexed.

Several writers confider the porpoife, as a fpecies of whale, and give it the name of *foufleur*. There are different kinds of them. Some of them have their backs of a dark grey, almoft black, and their bellies much lighter. Others are of a grey approaching nearly to white, from whence they have the name of *white porpoifes*. Thofe which we took, and whofe figure is reprefented in the plate, had their heads formed, not like the fnout of a hog, but almoft in the fhape of a bird's head, covered with a thick grey fkin, and the beak armed throughout with fharp white teeth like thofe of a pike. They had an opening (A.) on the top of their head, through which they fpouted water, and this was followed by a ftream of air attended with a noife fomething like the grunting of a hog. Their tail is horizontal, contrary to what is ufually found among other fifh, who have it perpendicular, when they are lying upon their bellies. It is of great ufe, no doubt, in affifting the porpoife to fpring out of the

water,

water, and to turn round in the air with ſo much eaſe, as I have mentioned in a former article; to effect which they only incline a little more on one ſide of their tail than on the other. From this poſition of their tail probably it is, that they derive that peculiar method of ſwimming, as if they were alternately riſing above the water and diving under it. The porpoiſe, which I am deſcribing here (and all thoſe we took were of the ſame kind) is, as I apprehend, of that ſpecies, which are called *moines de mer*. The fore part of the head terminates in a roll near the beginning of the ſnout or beak, anſwering to the border of the cowl. The back is blackiſh, and the belly of a grey, conſiſting of a pearl colour, ſomewhat inclining to yellow, interſperſed with black and iron-grey ſpots. It has three fins, curved and very thick; one on the back, the other two under the belly. Theſe, as well as the tail, are covered with a membrane, or thick coarſe ſkin, which being removed, five white cartilages appear, diſpoſed like fingers and articulated in phalanxes.

I diſſected the head and fins with an intention of preſerving them; but having hung them up over our cabbins near the flag ſtaff, ſome of our crew, in working the ſhip, inadvertently threw them overboard.

Porpoiſes almoſt always are found in ſhoals, ſwimming in a line, as if they were drawn up for an engagement. They ſeem to go in ſearch of the wind; for we remarked that in a ſhort time after they had paſſed us, the wind would riſe on that ſide, to which they directed their courſe. There is no fiſh perhaps, conſidering its ſize, that has ſo much ſtrength as the porpoiſe. Among thoſe, which we ſtruck, two or three diſengaged themſelves from the harpoon, either by tearing their backs or breaking the harpoon itſelf; although the ſpike was as thick as a man's thumb. Thoſe we took did always force the iron, and one of them twiſted it like the end of a ſcrew. There is a ſtrong ſmell attending this fiſh, as well as the ſhark, and it is ſo permanent, that after the diſſection I made, my hands were not free from it in three days, though I waſhed them very often with vinegar.

We

We had ſome of it ſerved up at dinner the day it was taken, which ſeveral others at the table, beſides myſelf, thought by no means ſo ill taſted, as it is generally ſaid to be.

Perceiving a change in the colour of the ſea, we thought proper to ſound; a precaution the more neceſſary in the latitudes we were in at this time, as there is very little dependence on the charts. Thoſe of Holland placing the coaſt of Brazil near 60 leagues more to the Eaſt than the French. Beſides, according to our reckoning and our obſervations of the ſun's altitude, we found ourſelves among, or at leaſt very near the ſhelves called Los Abrollhos, the extent and ſituation of which are not ſo exactly known and laid down in the charts, that they can ſafely be truſted to.

About half paſt ſeven in the evening, we ſounded with a hundred and thirty-five fathom of line, but found no bottom. Immediately after, another of thoſe birds, of which I have exhibited a figure, and which I take to be one of thoſe they call *tropical birds*, came and ſettled on the larboard of the quarter-deck. Here we endeavoured to catch him but he eſcaped. He then flew to the other end of the ſhip, and ſettling on the larboard of the fore-caſtle, a ſailor caught him in his hand. We put him in a hen coop, intending the next morning to faſten a ribband round his neck with this inſcription: *I was taken on the French frigate, Eagle, the* 20th *of November* 1763, *in* 16 deg. 44 min. lat. 35 deg. 10 min. long. *and was ſet at liberty the* 21ſt *in the morning*. At midnight we ſounded a ſecond time without finding any bottom.

Tueſday the 21ſt, at half paſt ſix in the morning, one of the mates being deſirous of examining the bird, which had been taken the evening before, and not holding him with ſufficient caution, our priſoner eſcaped, and deprived us of the pleaſure we propoſed to ourſelves in faſtening about him the ribband I have mentioned. From the time we had ſuffered the ſecond of theſe birds that we met with to fly away, we never failed to have one of them every evening about eight o'clock fluttering round our cabbins.

Having obſerved an alteration in the colour of the ſea all this day, we ſounded at eight in the evening. At the depth of 35 fathoms we found bottom, and brought up pieces of coral, ſhells, and rotten ſtone. At ten we ſounded again, and found 30 fathom with the ſame bottom. At midnight, no ſoundings.

At two in the morning of the 22d, ſounded again; 40 fathom, ſame bottom as before. At four, no ſoundings. The Abrollhos extend farther to the ſouthward than is marked in the French chart.

It is to be obſerved, that the author of admiral Anſon's voyage, being in the ſame latitude and longitude according to our reckoning, found the ſame ſoundings, which ſerved in ſome meaſure as a direction for us. This ſucceſſive difference, of ſoundings and no ſoundings, is the more remarkable, as by our reckoning we had not changed our courſe half a league; for from noon the day before we ſteered S. W. by the compaſs, till three quarters paſt ſeven, when we ſounded: after that S. $\frac{1}{4}$ S. W. till ten, then S. till midnight, when we came about again to S. S. W. after having made two leagues two thirds of way; at two we had ſoundings, and at four, ſteering the ſame courſe at the rate of five or five $\frac{1}{2}$ knots an hour, no ſoundings.

At noon we obſerved the ſun in our zenith, and could not miſs finding the altitude. Some minutes after we remarked that we had paſſed the ſun, and that our ſhadow lay ſouthward.

About three in the afternoon we made a ſignal to a ſail, we had had in view for ſome hours, thinking it to be our ſloop the Sphinx. She ſeemed to be making towards us, and was ſteering W. S. W. upon this we lay by for her; but finding that ſhe did not anſwer our ſignal, and having diſcovered that ſhe had but two maſts, we concluded her to be a negro ſnow going to Rio Janeiro. We kept S. W. before the wind till nine at night, when we changed our courſe to S. W. $\frac{1}{4}$ S. At midnight we ſounded without finding bottom.

At

At four the next morning, 23d of November, we ſteered S. W. ¼ W. and at ſix had ſight of the coaſt of Brazil, bearing W. and W. N. W. about fifteen leagues diſtance. At ſeven we came about to the wind in order to make the land; but the weather grew ſo hazy, that by ten o'clock we had loſt ſight of it. The colour of the ſea changing, we ſounded, and found a bottom of fine ſand at fifteen fathom depth. At eleven we ſounded again, and found the ſame bottom.

In meaſuring our diſtance upon the chart, according to our corrected longitude, we found ourſelves ſeventy leagues from the coaſt of Brazil, Eaſt and Weſt of the ſouthern point of the river Spirito Santo. At the ſame time we had ſight of land, and found ourſelves ſixty leagues farther Weſt than our reckoning; which confirms the remark of the author of admiral Anſon's voyage, that the tides ſet South Weſt. It is therefore very prudent not to truſt to theſe tides, nor to the charts, eſpecially the French ones, in the paſſage from the line to the river Plata.

Our firſt ſounding might probably have been taken on a bank of ſand in the open ſea, not marked out in the French chart, which is to be found in the Dutch chart of Wan Culen, marked *good bottom*, at 15 or 16 leagues off land. This is the ſame place where we ſounded at ten and eleven o'clock. The chart of Peter Goos is more accurate; and one of M. Buache is ſtill better.

After this the wind changing from N. N. E. to N. E. with a briſk gale, hazy weather, and a high ſea, we ſounded every quarter of an hour, and at three o'clock finding only nine fathom, we tacked, and put the Cape S. ¼ S. W. Our depth ſtill decreaſing, we came again to S. ¼ S. E. for half an hour; but finding that our water grew ſtill more ſhallow, though we were ſtanding out to ſea, we put about again, and brought the Cape to bear S. S. W. From this time the depth began gradually to increaſe, inſomuch, that at five o'clock we had twenty-five fathom water with the ſame ſandy bottom of the colour of bran, but ſomething more muddy than at the top of the bank. At eight we ſtood croſs it, in 35 fathom, with a bottom of white ſhining ſand. At

ten

ten we had forty fathom, with a bottom of rotten ſhells, and ſome coral.

Although this laſt ſounding had almoſt removed the apprehenſions we laboured under, through the error of the charts, in reſpect to the ſituation of the coaſt of Brazil, and the omiſſion of this ſand bank or ſhallow we had juſt met with, we thought it neceſſary to continue our ſoundings for the greater ſecurity. In conſequence of this reſolution, we found at midnight fifty fathom water, ſame bottom, but no coral: at four in the morning, being the 24th, ſixty fathom, ſame bottom as the laſt. From half paſt five we ſteered S. W. till noon. Theſe ſhallows are the flats of St. Thomas, which are very dangerous in ſtormy weather. They lie from ſixteen to ſeventeen leagues out at ſea, and the higheſt part of the ſhoal is not more than three or four fathoms below the ſurface of the water. Near the ſhore there is depth enough to paſs. The Portugueſe veſſels, which are employed in coaſting along theſe ſhores, and are well acquainted with them, keep between theſe flats and the land, but ſeveral of them have not been able to avoid running foul of the bank.

The ground between theſe flats and the ſhore is compoſed of ſand, which reſembles pounded glaſs, and that on the ſhoal itſelf is of rotten ſtone.

It is proper to obſerve, that the Dutch chart, of which I have ſpoken above, does not make the ſand bank, marked *good bottom*, extend ſo far as it really does, which is not leſs than through the 24th degree of latitude. I am not acquainted with the extent of it from Eaſt to Weſt. By our reckoning and obſervations of this day it appears, that the tides and the currents ſet ſouthward and weſtward. At ſix o'clock the night before, Cape St. Thomas bore nearly North Weſt of us by the compaſs, at the diſtance of fourteen or fifteen leagues.

The wind continued N. N. E. blowing hard, the ſky gloomy and covered. We ſteered with the Cape S. W. till ſix in the morning of the 25th. After that W. S. W. At ſeven, the evening before, we had ſounded, and found no bottom with fourſcore fathom of line.

On the 26th, from four in the morning to ſix, we were becalmed. We took this opportunity to ſound, but found no bottom. At ten in the evening we ſounded again, with the ſame ſucceſs. At midnight we found a bottom of grey ſand at the depth of 90 fathom.

At two the next morning (27th) our ſoundings were 85 fathom, with a bottom of grey ſand ſomewhat ſlimy.

At ſun ſet, though the horizon was not very clear, we ſaw land a head of us. We kept on our courſe, ſounding at ſeven in the evening, when we found thirty-five fathom, ſoft muddy ground, of a blackiſh grey colour, mixed with ſome ſmall ſhells. At midnight we ſounded again, at which time we had thirty-one fathom, ſame bottom.

The 28th at ſun riſe the land began to open upon us. We made towards it in order to take a view. At eight o'clock I diſcovered a ſmall iſland called in ſome charts *Aracari*. It bore N. W. 5 degrees W. of us by the compaſs, at the diſtance of about five or ſix leagues. The neareſt point to us, as far as I could diſtinguiſh at this time, was that which runs out fartheſt to the Eaſt on this ſide, and forms a peninſula. It bore W. $\frac{1}{4}$ N. W. of us by the compaſs, at the diſtance of about three leagues.

On the 29th, after having with much difficulty weathered the point of the iſland of *Gal*, and that of St. Catherine's, we came about four in the afternoon to anchor in ſix fathom water, muddy ground and very ſoft.

Marks of Anchorage.

Moored South South Eaſt, and North North Weſt.

In this bay, which forms a canal round the iſland of St. Catherine's, are three forts, and a battery of cannon near the entrance towards the town, on that ſide where we anchored. The firſt fort ſtands on the larboard ſide on coming into the bay. It is ſituated on a head-land in a ſmall iſland, called Parrot Iſland, N. E. $\frac{1}{4}$ E. and E. N. E. The name of it is *the Grand-Point-Fort.*

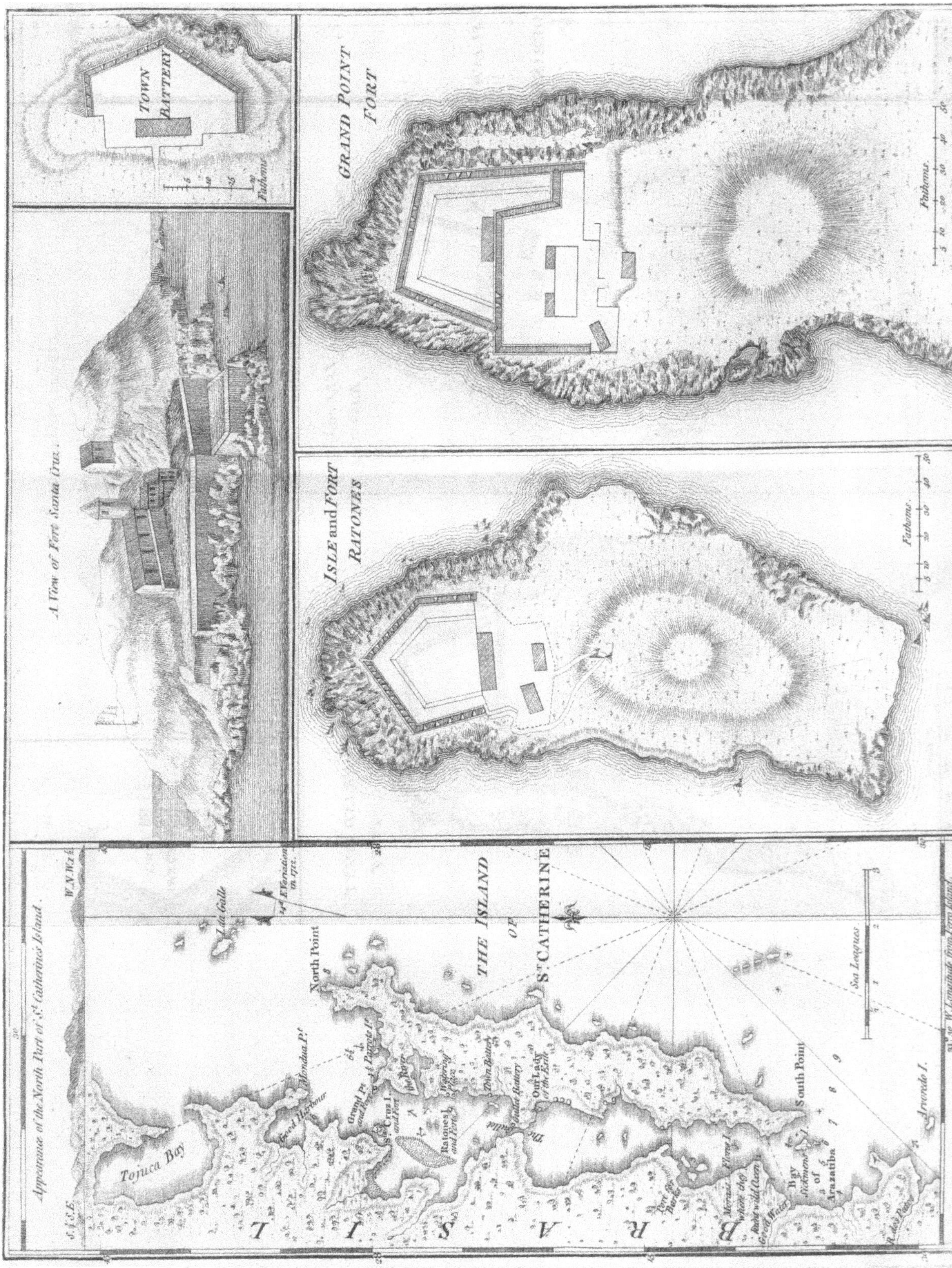

The material originally positioned here is too large for reproduction in this reissue. A PDF can be downloaded from the web address given on page iv of this book, by clicking on 'Resources Available'.

Fort. Almoſt oppoſite to this, but a little farther, is the ſecond fort, built likewiſe upon a little iſland near the main-land N. W. $\frac{1}{4}$ N. by the compaſs. This is called, *the fort of Santa-Cruz.* It makes a good appearance on entering the bay, being built on a terrace ſupported by arches. Here the commanding officer reſides. The third fort, which ſtands more in towards the town, is alſo ſituated on a ſmall iſland at almoſt an equal diſtance from the continent and the iſland, and goes by the name of *the fort of Ratonne.* Plans of all theſe forts are in the plate annexed. We lay at anchor in the middle of them, and the commanding officer gave us to underſtand by ſignals, that this was the beſt anchorage: but he had his reaſons for doing ſo, there being much more commodious anchorage farther in towards the main-land.

Upon our coming into the bay, we perceived the Portugueſe colours hoiſted at the top of ſome trees on an eminence in the iſland in the midſt of the woods, and placed ſo that it might be ſeen by the two advanced forts. They hoiſt this flag without doubt, as ſoon as they diſcover any ſhip at ſea, to give notice of it to the forts in the bay; for we ſaw it ſet up, and afterwards taken down again, as ſoon as we had come to an anchor and ſaluted the fort of Santa Cruz.

Before we moored, being over againſt this fort, which as well as the two others had hung out Portugueſe colours, we ſent our yawl with M. Alexander Guyot, our ſecond captain, who underſtands Portugueſe, to wait upon the commanding officer, and to aſk him, if upon our ſaluting the fort, he would return our ſalute, gun for gun. The Commandant ſent an officer of the garriſon back with M. Guyot to return the compliment, and to ſee who we were. As ſoon as they came on board, we caſt our anchors, and ſaluted the fort with nine guns, which was returned with the ſame number. The officer ſupped, and lay on board our ſhip that night, in order to conduct M. Guyot the next morning to wait upon the governor of this part of Brazil, who reſides at a little town, ſituated in the bottom of a creek in the iſland of St. Catherine's, five leagues South of our moorings, and to deſire his permiſſion, that we might wood and water. The ſame evening, the commanding officer of the fort of Santa Cruz, ſent us ſome refreſhments, and the next morning M. Guyot went with the Portugueſe Officer in the longboat.

The governor, whoſe name is Don Antonio Franciſco de Cardoſo y Menezes y Souza, colonel and knight of the order of Chriſt, and deſcended from a very illuſtrious family in Portugal, gave a very polite reception to M. Guyot, and granted every thing we deſired. Our longboat left the town about nine in the morning to return to us, but did not get back till ſeven in the evening, being detained by contrary winds.

As ſoon as the inhabitants on the coaſt perceived our frigate at anchor, three or four of them came along ſide of us in canoes, and brought us lemons, oranges, and ſome cabbages. But the commanding officer of the fort of Santa Cruz, obſerving this, ſent orders to all the huts forbidding the people to carry any thing on board our frigate, or even to come near it; they were neither to ſell us any thing or buy any thing of us. To inforce his orders more ſtrictly, he poſted ſoldiers in the neareſt huts, who were to watch the conduct of the inhabitants in this reſpect, and to prevent our ſtraggling about the parts adjacent. In the mean time he made us a thouſand proteſtations of his readineſs to oblige us, and was not at all ſparing of his civilities.

It is not to be doubted, that the Commandant immediately upon our arrival, diſpatched a canoe to the governor to give him intelligence of it. The next morning, while M. Guyot was gone to wait upon the governor, the Oviodore, or chief judge came on board our frigate to make a verbal proceſs of our anchorage, our force, and the reaſons which brought us hither. M. de Bougainville ſatisfied him in all theſe points, and he returned about noon. At his leaving the ſhip we ſaluted him with ſeven guns, which were immediately returned by the fort of Santa Cruz.

After

After dinner M. de Bougainville, accompanied by Meſſ. de Nerville, de Belcourt, and l'Huillier de la Serre, went to viſit the Commandant of this fort. Here they found a general officer of Rio Janeiro, who had been confined priſoner in it for four years, the commanding officer having received orders, never to ſuffer him to ſtir out of the gates. The crime laid to his charge was, that he had not punctually executed the orders of the court of Liſbon, in reſpect to the expulſion of the Jeſuits of Brazil, and had extended ſome favor to them. This gentleman had with him a Portugueſe, who acted as his ſteward and ſecretary; he was a man of good ſenſe and had been page to one of the Portugueſe ambaſſadors at Paris, where he had lived four years. The pleaſure of ſeeing Frenchmen again delighted him, and he was happy in ſerving as an interpreter to M. de Bougainville. His attachment to the impriſoned general had induced him to ſacrifice his liberty, and for the ſake of bearing him company he voluntarily partook of his confinement. This ſecretary accompanied M. de Bougainville, and the reſt on their return. In the account he gave us of the cauſes of the general's impriſonment, he exculpated him as much as he could, and told us even in the preſence of two officers, who had come on board with him, that he was indeed guilty of not having carried the orders of his court into execution as ſoon as he received them; but that the archbiſhop, who favored the Jeſuits had prevented it by giving him aſſurances that he had received counter orders; and that the other, as commanding officer, ought not to obey thoſe he had received, till they ſhould be confirmed. The execution therefore of them, whether out of reſpect to the archbiſhop or from other motives not known, was delayed too long, and the general was puniſhed for it by the loſs of his liberty. After this relation he begged of M. de Bougainville to take charge of a memorial in juſtification of the priſoner, and to deliver it to the Portugueſe ambaſſador in France upon our return, that it might be tranſmitted to the court of Liſbon. But no ſuch paper, I believe, ever came to the hands of M. de Bougainville.

When

When M. Alexander Guyot paid his viſit to the governor, he received an invitation from him, and was likewiſe deſired to deliver one on his part to M. de Bougainville and the Officers as well as the principal paſſengers on board our frigate to dine with the governor the next day, Thurſday the firſt of December.

As ſoon as it was light we ſet off, M. de Bougainville, de Nerville, de Belcourt, l' Huillier, Alex. Guyot and myſelf; and at half paſt one we arrived at the town, the name of which tranſlated into Erench is, *Notre Dame de l'Exil*, or *la Vierge Exilée, Our Lady in Exile.*

Almoſt all the officers of the garriſon came down to the ſhore to meet us. They received us at our landing with all the politeneſs imaginable, and conducted us to the governor's houſe through a large concourſe of people.

The governor met us at the outer gate, and led us into a large ſalloon, where we found dinner ſerved up. The governor, his ſon, who talked pretty good French, the town major, who knew enough of it to make himſelf underſtood, the Oviodore, two other officers and a friar of the Franciſcan order dined with us. Many other officers of the garriſon were in the room, but did not ſit down at table, and ſome of them waited upon us. Theſe military *waiters*, according to the account of an officer of the garriſon of Fort Santa Cruz, take that method to pay their court to the governor, who invites them all in their turns to dinner, and they wait upon one another.

The diſhes were dreſſed after the manner of the country, which is not very agreeable to a French palate. In eating their ſoup, which is a kind of ſolid glue, they do not make uſe of ſpoons, but eat it with the help of a fork. The bread was miſerable, brown, heavy dough, or rather paſte made in the ſhape of a roll, about three inches diameter, and an inch and a half in height. The outſide had ſcarcely felt the fire, and was only a little drier than the reſt. The inſide was ſolid, and looked like that buck-wheat flummery, which is the chief food of the inhabitants of Limoges, and is by them called *Gallette.*

The

The ſecond courſe was compoſed of a great number of diſhes, all dreſſed with ſugar, which is alſo an ingredient in moſt of their ſauces, as well as Carthamum, or the flower of baſtard ſaffron. The plates were pewter, not well ſcowered, and of an antique form. The covers were likewiſe very old faſhioned, but they were ſilver, and very heavy; ſo were the diſhes, and ſome drinking veſſels, which were of an octogonal cylindrical form, and about ſeven or eight inches high. Our liquor was at firſt brought to us in very ſmall glaſſes, ſuch as were formerly uſed for *liqueurs*, having a long ſtem, and the bowl ending in a point. But as it would have been neceſſary to drink one of theſe glaſſes at leaſt with every mouthful, I asked for a larger to mix wine and water in. Upon this they brought me one of the ſilver cylinders full. Another time I had a large goblet of cryſtal, the only one that appeared, and which held near a bottle, Paris meaſure. The other gueſts were ſerved in the ſame manner. Theſe great cups were handed from one to another, till they were empty. The wine we drank was port, and very good.

After dinner the cloth was removed, but we continued at table to drink coffee and to converſe. At this time I perceived that the Franciſcan had quitted the company. My deſign had been to addreſs myſelf to him in order to obtain ſome knowledge of the country and inhabitants; for though he knew nothing of the French language, nor I of the Portugueſe, I made myſelf ſure of being able to converſe with him in Latin. I acquainted the governor's ſon with my intention, and aſked him why the Franciſcan had retired. He told me, I might have perceived from the moment we came in, that the Friar had taken pains to avoid me, foreſeeing my deſign; for not underſtanding Latin, he choſe rather to keep at a diſtance from me, than be obliged to acknowledge his ignorance of that language. This want of learning, however, ſays he, is not peculiar to him; it is the caſe of almoſt all the eccleſiaſtics in the country.

The governor's ſon, from whom I received this intelligence is a captain in his father's regiment, and one of fifteen children

as-

as he informed u , that he had had, not by his lawful wife, for he never w s married, but by one or ſeveral miſtreſſes. The reſt of his children then alive were at Liſbon, where they enjoy according to the laws of that city the ſame honours and prerogatives as the legitimate children of nobility; baſtards being there, as I am informed, gentlemen by birth. One of the daughters of this governor is married to one of the miniſters of State in that court, and another of his ſons is in poſſeſſion of one of the firſt employments.

While the coffee was pouring, a dozen of the officers of the garriſon came in, and entertained us with a little concert of inſtrumental muſic; ſome playing on the German flute, ſome on the violin, one on the violoncello, and one on a ſpecies of hautbois. The ſame perſons played ſucceſſively on different inſtruments. There were beſides two blacks, who blew the French horn. The whole performance was very good. They had almoſt all the pieces of our beſt compoſers. The treble ſtring of their violins was made of ſilk.

After this we walked about the town, which appeared to me to contain about a hundred and fifty houſes, conſiſting of nothing but a ground floor with the roof over it. The garriſon occupies one part, and is compoſed chiefly of white men. The reſt of the inhabitants are almoſt all negroes or mulattoes; they are to be found of all ſhades from black to white. The greateſt number of both ſexes are mulattoes, and for the moſt part ill-ſhaped. They have beſides an air of ſavageneſs, ſuch as might be expected in a breed half Brazilian and half negro.

They go almoſt all of them with their feet naked, their head bare and very ill combed, a ſhirt, a pair of breeches, and ſome with a cloke, which they throw over their ſhoulder according to the Spaniſh faſhion. It is common enough to ſee their ſhirts and breeches in holes, and torn in more places than one; which makes the cloke very neceſſary. One meets with ſome, but they are undoubtedly of the richer ſort, who wear hats of a very large ſize, the brims being about ten inches in breadth, and flapped

7 down.

down. Thefe have their feet covered; they have likewife a waiftcoat, over which they throw an exceeding wide full cloak, which comes down to their fhoes: they fometimes throw a corner of it over the oppofite fhoulder, and that in fuch a manner as even to cover their faces. Inftead of a hat fome of them have a hood of the fame ftuff as the cloak, to which it is faftened, and with this they are ufed to cover their heads fo completely, that it is impoffible to know any perfon in that garb, unlefs by their walk, or fome other mark of diftinction, fuch as the colour of the cloak or the manner of wearing it.

The governor, and the officers, as well as the whole garrifon, wear cloth made up in the French fafhion: and indeed I was very much furprifed to fee officers in fo hot a country dreffed in cloth at leaft as coarfe as that of our foldiers.

The Oviodore, and officers of juftice are diftinguifhed by a large cane or ftaff, or by a fmall ftick, bent in form of a hoop, which the principal of them carry on their left arm above the elbow; the inferior ones have it faftened to the button hole of the left pocket of their coat.

The flaves go naked, except that the men have a pair of breeches or drawers, and fometimes a fhirt, that is fcarcely good for any thing: it often happens indeed that they have no other covering than a fimple piece of cloth about their fhoulders. It is very uncommon to fee any of them with a fhirt and waiftcoat. But when they have obtained their liberty, they are permitted to wear the doublet and cloak made of ftuff like the white people. The black female flaves are alfo naked, excepting a flip of linen which they tie round them by the two ends, and which reaches from their waift to the middle of their thighs at moft. When they are made free, they are dreffed like other women with a petticoat, and a fhift, the upper part of which is open before, a good deal in the manner of our fhirts: when they go out of doors, they put on a large piece of fine woollen, generally of a white colour, bordered with gold, filver, filk, or tape, according to the circumftances and condition of the wearer. It is about

K two

two ells in length, and one in breadth. It is put on ſo that one of the corners hangs down to the middle of the back, and in this reſpect bears a pretty near reſemblance to the cowl worn by our Carmelites. The oppoſite corner comes over the head, and the two others being drawn over the ſhoulders and arms down to the elbows, are brought acroſs each other upon the breaſt, like the mantlet of our French ladies. Sometimes too, inſtead of croſſing them upon the breaſt, they bring theſe ends under their arms on the ſame ſide, by which means their neck is left expoſed to view. This mode of dreſs is very inconvenient, and requires a continual attention to re-adjuſt it, ſometimes on the head, and ſometimes on the arms, the ſlighteſt motion of the body putting it out of order.

The Portugueſe women, who are eſtabliſhed or born on the iſland of St. Catherine's, and on the coaſt of the main land ſo far as our excurſions led us, are of a very fair complexion, notwithſtanding the heat of the climate. They have, generally ſpeaking, fine large eyes; but little colour in their cheeks. Both men and women for the moſt part lead a very lazy life, and truſt to their ſlaves the management of their houſhold affairs, and what little work is to be done out of doors. The land produces almoſt every kind of neceſſary, without their taking the trouble to cultivate it.

There are ſcarce any ſhops to be ſeen in the town. I found only two; one a lockſmith's, and the other an apothecary's. The negro women, who have obtained their liberty, carry fruit about in great baſkets on their heads, or ſquat down at the corners of ſtreets with their baſkets before them.

While we were agreeably engaged in attending to the muſic, his excellency the governor ordered a parrot to be fetched, which was exceedingly remarkable for its beautiful and variegated plumage. As he ſaw that we all of us admired it, he begged M. de Bougainville to accept of the parrot. Its whole plumage, particularly on the head, neck, back, and belly was adorned with feathers, ſome of a jonquil, others of a lemon colour, ſome carmine, others crimſon, and all intermixed with feathers of different ſhades of green, and of a lively blue, particularly at the ears. The governor told us, that this variety was partly owing to art, and partly to nature: for that when this bird is very young, and has ſcarcely more than the quills of the feathers riſing after the down, theſe quills are pulled out in different parts, and a kind of poiſonous liquid poured immediately into the place of them; that the feathers, which riſe afterwards where the quills have been pulled out, are yellow or red inſtead of green, which they would naturally be: but of a hundred birds, on which this operation is performed, there are ſcarce five or ſix, that do not die of it.

The governor added to this preſent fifty ſkins of toucans, ſtripped from the beak to the thighs, and dried with the feathers, which are partly lemon colour, partly carnation, and partly black, in croſs ſtreaks from one wing to the other.

He carried his generoſity ſo far, as to promiſe M. de Bougainville two guaras alive, a male and a female; and even offered to make him a preſent of thoſe which he ſhewed us, if there were no others to be got before our departure from the iſland. He could not however accompliſh his promiſe, a contrary wind preventing his return to the town the evening before we ſailed.

The guara is a bird of the ſize of a large French magpye. It has a long beak, which is crooked at the end; its thighs and feet are alſo long. The firſt feathers, with which it is covered after it is hatched, are black. This goes off inſenſibly, and becomes an aſh colour. When the bird begins to fly, all the feathers turn white; after which they become of a roſe colour, and growing more and more red every day, at laſt attain to the brighteſt ſcarlet, which they preſerve ever after. Though it is a bird of prey, feeding not only on fiſh but on all kinds of fleſh, which it uſually ſoaks firſt in water; yet it builds and lays its eggs on the roofs of houſes, and in holes of walls, as our ſparrows do. It always flies in company. The feathers, with which the ſavages adorn their heads, are taken from theſe birds. The two which the governor had promiſed to M. de Bougainville were juſt beginning to redden.

On Friday the 2d of December the Acadians with their wives, children and ſiſters-in-law, were ſet on ſhore, and lodged in a cottage on the continent, which had been aſſigned them by the Commandant of the fort of Santa Cruz. Here they were employed in waſhing the linen of the veſſel, and of ſeveral officers of the frigate.

The Commandant had pointed out to us a place near his fort to wood and water at. Some of the crew were ſent there for theſe purpoſes; but after ſeveral trials, they found much difficulty in getting this water, which ran from a little torrent. Beſide this, a ſmall whale having run a ground near the place ſome time before infected the air with ſo horrible a ſtench, that it was reſolved to aſk leave of the governor to water on the iſland. Our requeſt was granted with the greateſt politeneſs, his excellency giving us at the ſame time permiſſion to fiſh, ſport, and to go wherever we pleaſed. One of the inhabitants ſhewed us a ſpring near his houſe which formed a little rivulet, where Admiral Anſon had watered, and an oven built a few paces from it about ſeven or eight years before by ſome Frenchmen, who had put into the harbour. The water of this ſpring is very good; and we laid in a large ſtock of it. As to our wood, we got that from the place which had been mentioned to us on the continent, it being exceedingly commodious for that purpoſe: for after cutting down the wood on the brow of the hill, it was very eaſy to roll it down the ſide cloſe to the water edge, and ſo load our boats with it. What we cut was moſtly cedar, ſaſſafras, cinnamon, and Brazil wood, which is uſed in dying. There was very little of any other kind in this place.

The

The next day (Saturday the 3d) we went on ſhore upon the continent, and walked along the coaſt with an intention of ſporting. We found two or three ſoldiers poſted by the Commandant in the next cottage to that of our Acadians, who endeavoured to prevent us from going forward, and told us they had orders from the governor to do ſo. We pretended not to underſtand what they ſaid to us, as they ſpoke in Portugueſe, and continued our route, upon which they made no reſiſtance. In fact, it was not the governor, but the Commandant of the fort of Santa Cruz, who had given theſe orders. We proceeded above a league along the new road, which they are making to go by land to Rio Janeiro, and by the ſide of a chain of woods which cover all the eminences. Theſe woods are ſo thick, that it is not poſſible for any thing but wild beaſts and ſerpents to penetrate into them. We killed ſome toucans, parrots, tiepirangas, and one dove.

Sunday the 4th, M. de Bougainville accompanied by four or five officers went to dine with the governor, who had given us an invitation the Thurſday before. I remained on board to ſay maſs to the crew. The gentlemen who went were received and entertained ſplendidly as before. The wind and tide being againſt them hindered their returning to the ſhip that day, notwithſtanding all their endeavours to accompliſh it; they reſolved therefore to go back again to the town. The governor had apprized them of the impoſſibility of their getting on board at that time, and had done every thing in his power to engage them to ſtay. He repreſented to them, the riſque they muſt run by expoſing themſelves in ſuch a manner to the dangers of a channel full of ſhallows, banks of ſand, and rocks, which they would have the greateſt difficulty in the world to avoid, if the night ſhould come upon them before they reached the ſhip; that ſuch an accident would out it entirely out of their power to diſcover the *ſea marks*, and conſequently to keep the channel. Beſides, he had intended to give them the pleaſure of a ball, and had already invited ſeveral ladies, wives to officers of the garriſon.

The party being broke up by the departure of our gentlemen, he ſent to all the perſons who had been invited, to prevent their coming. But as ſoon as he underſtood that M. de Bougainville, and his companions were coming back to the town, he ſent out to meet them, and without acquainting them of his intentions, diſpatched freſh invitations for the ſupper and the ball, which was to follow.

After ſupper was over, at which the ladies were not preſent, the governor, without giving our gentlemen the leaſt hint of what they were to expect, engaged them only to go and paſs a few hours at the houſe of an officer of the garriſon, where, as he ſaid, they would find a very agreeable company. M. de Bougainville and the reſt conſented at firſt merely out of complaiſance, but they were agreeably ſurpriſed to find there ſeveral ladies, by whom they were perfectly well received. They had never imagined from the reputation the Portugueſe have of being extremely ſuſceptible of jealouſy, that they would have permitted their women to appear in ſuch aſſemblies. They ſtruck up however a kind of dance, in which the ladies figured as well as the gentlemen, and about two or three in the morning they retired very well ſatisfied with each other.

At this interview, M. de Bougainville took an opportunity of complaining to the governor of the behaviour of the Commandant of Santa Cruz, and obtained a general permiſſion to take whatever meaſures he ſhould judge proper for ſporting, fiſhing, wooding, and watering wherever we pleaſed. At parting M. de Bougainville invited the governor, with the Oviodore, and ſuch officers as he ſhould think fit to bring with him, to dine on board our ſhip.

In conſequence of this permiſſion we ſent our yawl out to fiſh almoſt every day, and ſhe conſtantly came back loaded with fiſh of many kinds, and in ſuch abundance as to ſerve the whole crew. The figures of them may be ſeen in the plates.

We went alſo every day a ſhooting either on the continent or in the iſland; though we ſoon left off going to the former, as we we found ſcarce any thing there but parrots, toucans, and ſome doves. In the iſland, beſide theſe birds, which I have mentioned, there were ſea larks, plovers, ſnipes, and ſome others in good number. The officers of the garriſon, and the people of the country, whether from want of courage, or from indolence, or from the danger of meeting with wild beaſts and ſerpents, which are in great plenty among the woods and moraſſes, never go out a ſporting, and adviſed us to follow their example. It muſt be confeſſed indeed, that the cottagers are not provided with arms; and the few which one meets with are old, the greateſt part made after the old faſhion, with wheels for the trigger, and very bad. They had ſcarcely even powder or ball.

Leſs timid than they, and to ſay the truth better armed, we more than once penetrated into the acceſſible parts of the iſland. By the aſſiſtance of our light boots we ſurmounted the obſtacles which preſented themſelves to us among the woods and thickets, from a ſpecies of thorny aloes, of which they are full. We never went alone, but always two or three in a company in order to aſſiſt each other in caſe of an attack from any overgrown ſerpent, or wild beaſt, particularly ounces; ſome claws of which we had ſeen here in the hands of ſome of the inhabitants mounted in ſilver, and which, as we were informed by them, were very common, and were more ravenous than even tygers.

One day when we were out in ſearch of game upon the iſland, and had ſeparated into different parties, I, with M. de Belcourt and his ſervant, kept along the ſide of a creek, which runs a conſiderable way within the land, and was called by us *the river*. M. de Belcourt amuſed himſelf with ſhooting at water-fowl. As we advanced along the border of this creek I perceived on the ſand recent traces of ſome four-footed animal, which, to judge by the marks, muſt be very large, and ſeemed to be a tyger. We followed theſe traces till we came to a very marſhy ſpot, where we did not dare to venture ourſelves, not knowing either the bottom or the extent of it. Returning by the ſame

way

way that we came, I perceived M. de Belcourt advancing towards us, and ſhewed him the traces.

Theſe, ſaid he to me, muſt certainly be the traces of a beaſt I ſaw hereabouts but juſt now, at the very inſtant he was darting into the thickets. He is about the heighth of the largeſt ſized Daniſh dog, and of a greyiſh colour. He went in at that place, let us purſue him. With all my heart, replied I. We made our way as well as we could among theſe marſhy thickets, which were ſo choaked up with a ſharp ſpecies of aloes, whoſe leaves are ſometimes not leſs than five feet in height, that we had all the trouble in the world to diſengage ourſelves from them. We beat about in vain for near two hours without ſeeing any thing of the beaſt we were in purſuit of: we only got ſight of the hind part of another, the hair of which ſeemed to be of a greeniſh grey: his height about that of the largeſt kind of ſpaniel: his tail ſeemed to be as green as the leaves of the plants, which ſurrounded it, and to reſemble that of a fox in thickneſs and in length. He hid himſelf among the buſhes at the inſtant M. de Belcourt was going to fire at him.

The heat was now ſuffocating. We ſtopped and ſat down on the ends of ſome branches, leaning our backs againſt a tree. We had with us ſome oranges and ſome ſea biſcuit. While we were regaling ourſelves in this poſture, we were ſtunned with the inceſſant hiſſings of ſerpents, which ſurrounded us, and reduced us to the neceſſity of keeping conſtantly upon our guard with drawn ſabres. After this breakfaſt, of which we had ſtood in great need, we continued our ſport, drawing towards an eminence, at the top of which we diſcovered a cottage. Being arrived here, we found M. de Bougainville and his ſervant. Two Portugueſe women, whoſe figures were not very inviting, received us, and diſplayed for near two hours, that we continued with them, the utmoſt freedom in their air and converſation. They had a tame parrot, which was tolerably pretty, and talked well. We propoſed to purchaſe this bird; but the women refuſed to part with it. They longed for every thing we had,

handkerchiefs,

handkerchiefs, knives, hats, and even our fuzils and ſabres: they aſked us for all theſe without ceremony, and if we had been willing to give into their way of thinking, neither our cloaths nor even our ſhirts would have incommoded us on our return. We contented ourſelves with a few oranges, and ſet out to dine on board. On entering the wood we ſeparated again from M. de Bougainville and his ſervant without intending it. The path which M. de Belcourt and I took led us to a moraſs, where the trees were luckily at a ſufficient diſtance from each other: Here we ſaw ſeveral ſerpents about the thickneſs of the ſmall of a man's leg, and others leſs, ſome of a reddiſh colour, others red and yellow, and others grey, which laſt a good deal reſembled adders of the largeſt ſize; but inſtead of attacking, they fled before us. When we were almoſt got to our boat, M. de Belcourt fired at a bird, called *the ſpoon-bill*, and broke only one of his wings. He took it up and brought it on board. It was a young one, and all its plumage was of a faint roſe colour; the proceſs of the quill from which the beards of the wing feathers riſe, was of a bright roſe colour. Its legs were a foot long including the thighs, and of a light grey as well as the feet, which were webbed, like thoſe of geeſe. Its beak was ſix inches in length, and both the upper and under part flat, grey towards the root, and white towards the extremity: it began to ſpread at about two thirds of its length, and ended in a ſpatula, of two inches and a half diameter in its greateſt breadth. We carried it to the ſhip, where it lived three days on ſome ſmall fiſhes, and bits of freſh meat, which were forced down its throat; for it would not eat of itſelf. When any one came near it, it made a noiſe with its bill as loud as that of two wooden battledores ſtruck one againſt the other.

Some of our ſailors gave it the name of flamingo, but that of *ſpoon-bill*, or *palette (battledore)* is more ſuitable, on account of the ſhape of its bill, very unlike to that of the flamingo, which is made almoſt in the commoneſt form of a bird-bill.

The next day we went again on the iſland in purſuit of the animal we had not been able to meet with the evening before. There were ſeveral that made their way into the ſame wood. M. l Huillier found there another beaſt of the ſame ſize as that I have mentioned, but of the yellow colour, and nearly of the ſhape, of a lion. He fired three times at him loaded with bullet, two of which wounded without ſtopping him, and without making him go one ſtep either faſter or ſlower. We followed him by the traces of the blood, but he ruſhed in among the thickets, and we ſaw no more of him. The heat being very intenſe, and we having ſcarcely more than time ſufficient to get on board for dinner, contented ourſelves with killing parrots, plovers, ſnipes and ſome, other birds. As I was curious to have a humming bird, of which I ſaw many flying round our heads, and could not, or at leaſt did not know how to take them alive, I ventured to fire at one, which was fluttering about like a butterfly, and hovered in the ſame manner over a ſmall branch of a tree. The little bird, whether through fear or the violent concuſſion of the air, dropped inſtantly. After having looked for it a long time, I found it at laſt dead on a leaf of the ſame branch. The figure of it in its natural ſize is to be ſeen in the plate.

Some call this bird, *Liſongere or Beequeſleurs*, becauſe it is continually fluttering about flowers, like the butterfly, and ſucks the moiſture of them in the ſame manner. The whole compaſs of its body with the feathers is not larger than a common nut. It has a tail near three times as long as its body; its neck is rather ſmall, its head in proportion, and its eyes are very ſharp. The bill is ſomewhat whitiſh at the root, the reſt of it is black: it is as long as the body of the bird, is ſmall and very ſharp. The wings are long, thin, and very extenſive in proportion; the extremity of the feathers reaches to two-thirds or thereabouts of the tail, which as well as the wings is of a purpleiſh brown. The reſt of the plumage is green with a gold caſt, as if one had ſpread a layer of green almoſt tranſparent over a leaf of gold.

The

The neck and head are of a deep blue, gilt in the ſame manner. Theſe colours vary according as the light ſtrikes more or leſs forcibly on the different parts. Sometimes the whole plumage of this bird reſembles a pigeon's neck, or the green feathers on the wings of wild ducks; ſometimes it is of a fine blue, ſometimes of a fine green, ſometimes of a purple, and all theſe mixed with the ſplendor of a lively gold colour, bright and burniſhed. The tongue of this bird in miniature is forked, and has the appearance of two twiſts of red ſilk. Its feet are ſhort, black, and furniſhed with very long claws.

There are ſeveral ſpecies of them, which differ both in ſize and colour. One of the ſmall kind, which I have preſerved in brandy, has white feathers from the breaſt to the tail. The colour of the reſt of its plumage is like that of the others.

The female lays but two eggs, of the ſize of a ſmall pea. They build their neſts in orange trees with the ſmalleſt ſtraws they can find. The Portugueſe, who lived in the cottage near which we watered, gave us one of theſe neſts with two young ones in it, which were not yet covered with the firſt down. He had juſt taken it with the father and mother, cloſe by his habitation; we put it down on a ſtone bench at the door of the houſe, while we were eating an orange, and had ſcarcely turned our backs, when a cat came and carried off both neſt and young ones. Theſe neſts are of an admirable conſtruction, and about the ſize of a half crown. The Brazilians call this bird by the names of *Guainumbi*, *Guinambi*, *Aratica*, *Aratarataguacu*. The Portugueſe call it *Pegafrol*.

We had beſide theſe a third kind, ſomewhat larger than thoſe I have been deſcribing, but much leſs than the ſmalleſt wrens we have in Europe. The feathers of their head begin towards the middle of their upper bill. They are exceedingly ſmall at their riſe, are diſpoſed in ſcales, and grow larger as they are nearer to the head, at the top of which they form a little tuft of uncommon beauty for the brilliancy of the gold, and the variety of colours, which change according to the direction of the rays

 of

of light, or the poſition of the ſpectator's eye. Sometimes the plumage of this bird is of a black equal to that of the fineſt black velvet, ſometimes of a pea green, ſometimes yellow. At other times it reſembles cloth of gold ſhaded with all theſe colours. The back is of a dark green ſhot with gold. The large feathers of the wing are of a deep violet, approaching ſometimes to purple. The tail is compoſed of nine feathers as long as the whole body, and of a black mixed with brown, purple, and violet, which form a moſt agreeable aſſemblage of colours, and have the ſame changeable property as above. The whole lower part of the belly likewiſe exhibits a mixture of black, violet, green, and yellow, which always ſtrike the eye of the obſerver differently, according to the difference of his own ſituation or of that of the bird. Its eyes are of a lively, brilliant black, not inferior to the fineſt poliſhed jet; its legs are ſhort, and black, as well as its feet, which are compoſed of four claws, three of them in front, all furniſhed with black, crooked, ſharp talons, very long in proportion to the reſt of the body. When it flies, it makes a buzzing with its wings, very much like that of certain large flies which we ſee in France fluttering about among the flowers. It builds its neſt upon ſhrubs, among high branches of orange, or other low trees. In our French iſlands it goes by the name of *Colibris*, and ſometimes *Quinde*. The Spaniards call them *Tomineios*, becauſe the neſt and bird together do not weigh more than the Spaniſh *Tomin*.

When we came on board, we found a Spaniard there, who was ſettled in the country, and to whoſe care we had, the day we came to anchor, committed a ſheep that was diſtempered and very lean: the ſheep was to graze about the man's cottage, by which means we hoped to reſtore him. This Spaniard had brought us ſome hundreds of oyſters. They were much larger than the white oyſters of Saintonge; for the ſhells were at leaſt five inches in diameter. We do not eat fatter or better oyſters in France. They were a perfect cream, both in taſte and whiteneſs. We did every thing in our power to induce the Spaniard to diſcover the place where he found them, but we could not ſucceed.

ſucceed. All we could obtain from him was a promiſe of bringing us ſome more, and this was not performed till a day or two before our departure. We endeavoured to find out his cottage, but in vain; he had not given us a proper direction; and as for our ſheep we ſaw no more of it: he made us ample amends, however, when we were on the point of quitting the harbour, by a preſent he brought us of ſome thouſands of oranges and lemons, and of ſeven or eight hundred of the ſame kind of oyſters.

There was a Portugueſe too on board, who had brought a large handſome canoe, which our captain M. Duclos Guyot had bought for M. de Bougainville, who imagined at that time, that it might be very uſeful to us at the Malouine iſlands, to which we were bound. She was fitted out with her * *Pagaïes*, or PAGALLES, as father Labat calls them in his *Nouveaux Voyages*. This canoe was made of a ſingle trunk of a cinnamon tree hollowed, nineteen feet, odd inches in length, and three feet in breadth on the inſide, and about the ſame depth. Some of our officers, as well land as ſea, who had been in Canada, underſtood the working of her. She was uſed on our fiſhing expeditions. But when we put into Montevideo, M. de Bougainville parted with her to a Spaniſh officer for eight piaſtres; ſhe had coſt him about eighteen French livres.

About four o'clock in the afternoon we went over to the main land, and viſited ſeveral cottages on the coaſt, where we made proviſion of lemons, oranges, and ſome pine-apples, which we found ripe. This fruit, and the plant which bears it are known at preſent in Europe, as they bear the voyage very well, but there is a very great difference both in flavour and ſmell between the ſpecimens of this fruit produced in France, even in Provence and Languedoc, and what is found in Brazil. It grows there of itſelf without cultivation, and in great abundance. It turns the knives, with which it is cut, black, and ſpoils them; which pro-

* Paddles.

bably

bably has given occaſion to ſome authors to ſay, that the rind of it is ſo hard, that it blunts the edge of a knife. It is true, that if, after cutting the pine-apple in ſlices, you neglect to waſh and wipe your knife very well, you will find it after a few hours ſpoiled and ruſty, as if you had put *aqua fortis* ſomewhat lowered upon it. The juice of this fruit is of great efficacy in taking ſpots out of cloaths. That of Brazil, they will tell you, is a preſervative againſt ſea-ſickneſs.

In my walk I gathered ſome ſeeds of plants, and ſome grenadillas, with a ſmall red fruit of the colour of cinnabar, which bears a pretty near reſemblance to the love-apple. A Portugueſe, who was with us, told me, they call it *Maracuja*, the figure of it is in the plate annexed.

The plant which bears this fruit is prickly, the leaf is very like that of the *Stramonium furioſum*, but not ſo large. Under the rind of the fruit is a pulp, of one ſixth part of an inch in depth, white, and of the conſiſtence of that of the Calville apple, of a ſweetiſh but inſipid flavour. The inſide is intirely filled with flat ſeeds, of the ſame form as thoſe of the large Pimento or long pepper. The Portugueſe informed me, that the fruit, Maracuja, was never eaten, altho' he did not know that it had any dangerous qualities.

The Grenadilla of Brazil is round, yet rather flat at the ends, and of the ſize of a pullet s egg. Its bark is very ſmooth, glittering on the outſide, and of a carnation colour, when the fruit is ripe. On the inſide it is white and ſoft, its thickneſs about the eighth part of an inch. The ſubſtance which it incloſes is viſcous, it is of a refreſhing and cordial nature, the taſte of it is between ſweet and ſour. It may be eaten in quantities without any inconvenience. There are to be found in it a number of ſmall ſeeds or kernels much reſembling linſeed in ſhape and not ſo hard as thoſe of the common pomegranate. This whole ſubſtance is ſeparated from the bark by a very thin ſkin. The plant which bears this fruit twines about the trees, and reſembles as to its leaves and flower, what we call the paſſion flower. It diffuſes a very ſweet ſcent. To eat the Grenadilla in perfection, it ſhould not be ſuf-

fered

fered to ripen entirely upon the plant. It would decay and dry up. It muſt be gathered a little before it is ripe, and kept a few days.

Since our arrival at Brazil we were conſtantly in ſearch of parrots, but could not find any tame ones to be bought. In the tour that we made, we had the good luck to meet with ſome complaiſant Portugueſe, who parted with one to Mr. l' Huillier; this officer found means alſo to get one for Mr. de Belcourt. Upon our returning on board, a Spaniard who talked a little French, and whom we had commiſſioned to procure us ſome, offered us four, two of which were already reared, and talked the Portugueſe language, as did the two of which I have already ſpoken. The other two were but juſt taken from their neſt, and could not feed themſelves. I gave a ſtriped ribband for one of theſe laſt; and I preferred it with an idea that it would learn the French language with greater facility. I kept him till the beginning of May, when he died of a catarrh in the head. This catarrh had cauſed his eyes to ſwell. It fell upon his lungs, and having rendered him aſtmatic, it was impoſſible for me to ſave him.

Among theſe parrots there were three kinds, which differed in their plumage and ſize. One of M. l' Huillier's had the feathers of his neck and ſtomach of a tawny and changeable red, mixed with a little grey; the top of the fore part of the head of a vermilion colour, rather faded and extinguiſhed, the tips of the wings of a brighter red than that of the roſe, and ſeveral of the feathers in the wings and tail of a fine carmine; others of a very fine azure blue, and ſome black: all the reſt of the body was green. He ſpoke Portugueſe extremely well, and learned French very eaſily. He died juſt upon our arrival at the Malouine iſlands. The ſecond was bigger than any we had: the top of his head was of a vermilion red, the two ſides of a light blue towards the ears, and which grew fainter even ſo much as to become grey in proportion as the feathers were at a greater diſtance from them. The wings and tail were like thoſe of the firſt. The others were ſcarce above half that ſize; they reſembled them however as to their plumage, except that the red upon their heads was much more lively, which might perhaps be owing to their being younger. Monſ. de Bougainville's parrot died of the ſame diſorder as mine, during our ſtay at the Malouine iſlands; M. de Belcourt's fell into the ſea and was drowned, ſo that out of ſeven, we brought no more than two to France, Mr. l' Huillier's large one, which I delivered to him ſafe and ſound at Verſailles, and one of the ſmaller kind which had no tail, for he plucked out the feathers of it as faſt as they grew. The ſailor to whom it belonged, had not taken near the ſame care of him as we had done of ours, and yet preſerved him. It was impoſſible to ſpeak better than he did; and he imitated the cries of the children we had on board, and thoſe of the cabbin boys when they were whipt for any fault they had committed, the cackling of the hens, and the noiſe of all the other animals we had in the frigate, ſo well, as to deceive every body that heard him.

Paſſing by the habitation in which we had lodged our Acadian families, we heard a noiſe like that of a wood-cutter felling of wood. We aſked a freed negro, what it was? It is, anſwered he, a monkey that ranges about the garden to eat the fruit and the corn, and is giving notice to his comrades to come and aſſiſt him; but if I had a good gun like yours, I would ſoon diſlodge him. He has been two or three days making this racket. One of our boatſwains lent him his gun; the negro loaded it with large ſhot, followed the noiſe, and ſhot at the monkey twice without making him run away: at the third ſhot he fell dead at the foot of the tree. The boatſwain brought the monkey on board the frigate where we had opportunity to examine him at our leiſure. He was near two feet eight inches high, when ſtanding upon his hind legs; his hair was long, and of a fawn coloured brown all over his body except under the belly, which approached the clear fawn colour. His brown beard began from his ears and fell near five inches upon his breaſt; his feet and hands were black; his ears, deſtitute of hair, were well detached from each

each other and his face covered with a tawny down, ſo cloſe as to be hardly diſtinguiſhable from the ſkin. His eye brows were of a darker hue and prominent. His tail was as long as his body including his head.

I know not at what ſport he had loſt his left eye: this, however, was not to be perceived without a cloſe examination; for in the ſocket he had ſubſtituted a ball, compoſed of a gum which was unknown to us, of rotten wood and ſome very fine moſs, the whole mixed up together. The eye-lid covered this ball as if it had been really the globe of the eye. Whether he had contrived this falſe eye to appear leſs deformed, or to cure his wounded eye, or to defend it from the inſults of flies and other inſects, I leave to conjecture. We obſerved alſo, that this monkey appeared old, for the ſkin of his face was greatly wrinkled, and he had ſome white hairs in his beard. We ſaw but this one during our ſtay at the iſland of St. Catherine's, though we were told that there were a great number, and that the inhabitants eat the young ones, which are very good. They endeavoured even to perſuade me that one of the ragouts of which I ate at the Governor's, and which I took to be an excellent rabbit, was really a monkey. Be this as it would, many others ate of it as well as myſelf, and appeared well pleaſed with it.

The maſter of the habitation near which we got our water, having perceived that Mr. le Roy, lieutenant of our ſhip, had a great inclination for a pretty little bird that he had in a cage, and which ſang very well, made him a preſent of it. This bird is called in the Brazils, *Guranhé Engera.* It is of the ſize of a Canary bird. Its wings, back, neck and tail are blue, with ſome white ſpots about the middle of the large feathers of the wings and tail, diſpoſed in the ſame manner as theſe ſpots are in the wings and tail of the gold-finch. From the under part of the bill along the breaſt to the under part of the tail, all the feathers are of a golden yellow, bright and glittering; its warbling varies like that of the Canary, and it imitates the ſinging of other

M birds.

birds. There are ſeveral ſorts of them. The Brazilians call them alſo *Teitei*. For its Figure, ſee the Plate.

As I was walking in the fields with our captain, I perceived him gathering a large quantity of a plant with yellow flowers, which I took at the firſt glance for the yellow amaranth, which abounds on the riſing grounds on the coaſt of Terra Firma. Curioſity led me to aſk him the uſe which he meant to put it to. He told me, that it was the *Doradilla*; that when he was at Valparaſo, he had heard it called by that name; and that in that town, as well as in all the others in Peru, where he had been, they uſed a great deal of it in infuſion for the cure of pains in the ſtomach. Our captain was ſubject to them at times. I gathered a pretty large quantity of it, and we drank it ſometimes by way of tea. The taſte of it is agreeable enough. Others call it *Vira-verda*; this is the name that is given to it at Montevideo alſo. Frezier, in his account of his voyage to the South ſeas, ſays, that a French ſurgeon made uſe of it with great ſucceſs in the cure of the tertian ague. But the *Doradilla* which the Spaniards have, is a kind of ſpleen-wort, the leaf of which is curled. They attribute great virtues to it. The ſtalk and leaves of the vira-verda, which we ſpeak of at preſent, are ſpongy, and like the yellow amaranth, its flower is an aſſemblage of ſmall yellow buds, the leaves of which are pointed. The flowers of the amaranth are in form of a roſe, and the leaves of it are diſpoſed in the ſame manner.

At our return from fiſhing, abundantly ſupplied as uſual, we examined the different ſorts of fiſh, and among them found that which is called in the Brazils *Panapana*. The one that I give the figure of was two feet and a half long from the head to the beginning of the tail, the diſtance between the eyes was ten inches. Its ſkin was rough and hard like that of a ſhark, but conſiderably finer, nearly the ſame as that of the ſkin of a kind of ſhark, commonly ſlim and of a middling ſize, which our ſailors call *Demoiſelle*; we catched three or four of them during

our

our ſtay at the iſland of St. Catherine's, and two at the Maldonades at the mouth of Rio de la Plata.

The head of the Panapana is flat, ill-formed, and in the ſhape of a hammer. Its eyes are at a great diſtance from each other; being placed at the oppoſite extremities of the head. Its mouth and tail are like thoſe of the ſhark, its teeth are very ſharp; but I did not find ſeven rows of them. Our ſailors gave it the name of *marteau* or hammer, which is very applicable to its ſhape.

Among the number of plants which I gathered, was a ſort of pepper, or pimento, very common in the fields along the ſkirts of the woods. Its taſte is infinitely ſharper than that of the pimento or long pepper, which we are accuſtomed to in France. On this account our ſailors called it *piment enragé*. This fruit is of the ſame length form and colour, but at leaſt twice as thick as that of the Barberry tree. It is at firſt green, and grows red as it becomes ripe. The flower which precedes it, is like that of the pimento. The plant which bears it, grows to the height of about two feet. It is full of branches and joints; its ſtem is round, green, and rather ſlender. The leaves of it are in ſhape like thoſe of the *ſolanum hortenſe*, or garden nightſhade; but as ſmall as thoſe of the *chenopodium fœtidum* or *vulvaria*, which they reſemble much. One of the ſmall fruits of the *piment enragé* put into ſauce, heightens the flavour as much as an entire one of the larger ſort. This induced our ſailors to lay in a large proviſion of them.

I had alſo furniſhed myſelf with all the ripe ſeeds of the plants which I found, and having met with ſome Portugueſe women in a hut, who were picking cotton to ſeparate it from its ſeeds, they gave me a handful of them. They did me the greater pleaſure, as I was very deſirous of having ſome, and as I could not gather any from the plant, the ſhrub being juſt then in flower. The wood of it is tender and ſpongy; the bark thin and grey. Its leaves are of a bright green when young, but grow of a deeper colour as they approach to maturity, or as the ſhrub grows old. They are large, and divided into five parts, which terminate in a point. Thofe which are neareft the flower are only in three parts, and refemble much thofe of the *Ricinus*. Its flowers are almoft like thofe of a fmall fhrub, which is now in fafhion for the decoration of our parterres, called *Althææ*; They are not however quite fo open. They are yellow towards the end, and fpotted with red at the bottom. It is a pentapetalous plant, fupported by a calix of fmall green leaves, which are hard and pointed. To the piftil fucceeds a bud, or oval fruit, which in its ftate of maturity is of the bignefs of a duck's egg. This fruit is divided into three or four different cells, filled with a white ftringy fubftance, which inclofes ten or twelve feeds of a dark brown colour fticking together two by two, like wheat in the ear. Thefe feeds are of the fize of a pea, and about three or four lines in length.

This ftringy fubftance is that which we know by the name of cotton. It fwells and fprings in the fhell, which contains it, in fuch a manner as to force it open when the fruit is ripe. At that time the feeds, full of an oily fubftance, feparate themfelves, with the locks of cotton which inclofe them, and fall from the fruit, unlefs care is taken to gather them in time.

The Portuguefe muft certainly be unacquainted with the machines which are made ufe of in our Antilles, for feparating the cotton from the feeds which it inclofes, and to which it fticks; or elfe the Portuguefe women I have feen employed at this work, did it merely by way of amufement; for they feparated it bit by bit, only by pinching the cotton between their fingers. They fpin it afterwards to make cloth of it; but I do not know with what machine, as I never faw them at that work.

This is the only kind of cotton tree which I found cultivated in the ifland of St. Catherine's, and upon the coafts of the Continent in its neighbourhood. It is very different from the cotton tree in the Brazils, of which Dampier fpeaks in the following terms: " Its flower is compofed of fmall filaments al-
" moft as loofe as hair, three or four inches long, and of a dark
" red except the tips, which are of an afh colour. At the bot-
" tom

" tom of the ſtalk there are five narrow ſtiff leaves about ſix " inches long." The kind which Frezier ſpeaks of, reſembles in every reſpect that which I have deſcribed, except that the ſeeds of the latter are not ſeparated from each other, and diſperſed in the cotton, as that author ſays, and as he has repreſented them in the figure which he has given of the fruit. It ſeems as if father Labat had copied from this figure of the cotton tree the one which he has inſerted in the ſecond volume of his new voyage to the American iſlands, or perhaps Frezier may have taken it from him. The figures given by each of them are exactly alike.

In a hut a little farther on, where we went to beg ſome water to drink, the woman who gave us ſome was employed in ſtripping leaves, with long thin thorns on the ſtalk, from a kind of reed very common along the ſides of the woods and roads. She drew from them a ſort of green thread extremely fine, much like raw-ſilk, and of a light green colour. She told us, that ſhe afterwards ſpun this ſtringy ſubſtance to make lines and fiſhing nets of it, which ſhe ſaid laſted a long time. Perhaps it might be made uſe of for other purpoſes.

Not far from thence, I ſaw for the firſt time, a kind of aloes called *pithe*, the leaf of which when ſteeped like hemp affords a ſubſtance fit for ſpinning, and of which they make linen cloth in the Eaſt. From the midſt of a ſcore of leaves, about five feet high, and at leaſt three inches thick at the bottom, their edges thorny, ending in a point, hollowed out, and of a fine green, ſprang up a green ſtalk of about eight inches diameter at the bottom, which diminiſhed gradually to the top, and grew to the heighth of at leaſt thirty feet. From the heighth of about twenty feet of this ſtalk quite to its ſummit, there ſprang branches to the number of twelve or fifteen, adorned with a number of ſprigs, almoſt like the growing ſtalk of the lilly plant, when riſing about two inches from the earth. Theſe tufts of ſprigs grow irregularly along the branches, which are deſtitute of any other foliage, and ſpread themſelves almoſt horizontally. Without doubt, theſe ſprigs when arrived to a certain pitch of maturity, break of themſelves, and take root in the earth where they happen to fall. I picked up about fifteen with their roots, which I carried on board, where we planted them in boxes placed over our cabbins. They did very well there, and we ſhould in all probability have preſerved the greateſt part of them, if, notwithſtanding all our care, two cats which we had on board had not ſcratched up the earth of theſe boxes, and poiſoned it with their urine and excrement. We reſolved at laſt, though rather late, to cover them with fiſhing nets, ſupported by hoops, and we preſerved two plants of them as well as ſome cotton trees, raiſed from the ſeeds which we had ſown. They were all tranſplanted at our arrival at St. Malo's, into the garden of Beau Sejour at St. Servant, which was occupied by M. Duclos Guyot our captain.

The Portugueſe have perhaps remarked that the ſprigs of the *pithe*, which have thus taken root of themſelves, do not thrive ſo well as thoſe, which have been carefully put into ground well tilled. This is probably what induces them to make holes of about a foot ſquare in the earth beneath the branches, and about the plant itſelf, where I found five or ſix of theſe ſprigs planted; and which, in reality, ſeemed to have thriven better than thoſe which had been wholly abandoned to nature. I cannot ſay, whether the *pithe* bears any other fruit, or whether it multiplies by any other means.

Beſides lemons and oranges, there is in the iſle of St. Catherine's a ſort of refreſhment for ſportſmen. This fruit, which is very common, is called the American Indian fig. It is in ſhape much like our figs. Its firſt ſkin is green; it then grows rather yellow, and afterwards aſſumes the colour of red lacker on the ſide which has been expoſed to the rays of the ſun. This ſkin is ſtuck full of very ſmall prickles. Thoſe who gather this fruit and peal it muſt be very dextrous, not to fill their fingers with theſe prickles, which are almoſt imperceptible. Happily they cauſe more uneaſineſs than miſchief, till one has found the method of getting rid of them.

Under

Under the ſkin, which is about as thick as that of a fig is found a white covering, thin and tenderer than the other. It incloſes a ſoft ſubſtance, of a bright red, mixed with ſmall grains like thoſe of the fig. This ſubſtance has an acid taſte, a little ſweet, and extremely grateful. When people eat a conſiderable quantity of it their urine becomes red, but without any harm reſulting from it. This fruit is even cooling. Our captain, his two ſons and myſelf, were almoſt the only people who eat of it: the others did not dare to follow our example through apprehenſion of ſuffering ſome inconvenience from it.

In order to avoid running any of the riſques which I have mentioned in gathering theſe prickly figs, get a ſmall piece of wood ſhaped like a bodkin, and ſtick it into the fig near the ſtalk; cut off this ſtalk with a knife, and holding the fruit in this manner at the end of the bit of wood, peal it lightly all round, without touching it with your fingers.

On Saturday the tenth of December we ſent out ſome people to ſhoot in the iſland, in order to procure ſome game to treat the Governor, whom M. de Bougainville had invited to dine on board our frigate the next day. They brought nothing but parrots, ſnipes, and ſome other birds.

We found in a wet marſhy ſoil a prodigious quantity of a ſort of crab which live on ſhore, and make their retreat in holes which they dig. They give them the name of *tourlourous*; the biggeſt are not above two inches wide. The ſhape of their ſhell is almoſt ſquare, of a browniſh red, growing lighter by degrees towards the belly, which is of a clear red. This ſhell or helmet is pretty ſtrong though thin. Their eyes are of a ſhining black and as hard as horn. They ſhoot out and drawn in again like thoſe of lobſters.

Theſe crabs have four legs on each ſide, each of them compoſed of four joints, the laſt of which is flat and terminates in a point. They make uſe of them to walk ſideways, like common crabs, and to dig up the earth. Beſide theſe, they have two other

other legs or claws bigger than the former, but eſpecially the right, which is at leaſt double the ſize of the other. Theſe claws or pincers are of a bright red, ſhaped like thoſe of ſea-crabs; they make uſe of them for cutting leaves, and the roots of plants on which they feed. When they ſee any thing which frightens them they ſtrike theſe two claws againſt each other, as it were to frighten their enemy, and lift up the biggeſt of them perpendicularly, marching thus in a ſtate of defence; but retiring at the ſame time into their holes. Theſe claws as well as their legs, are ſo ſlightly fixed to their body that they come off in the hands of thoſe who endeavour to take them, and the tourlourou eſcapes.

Both ſexes have their tails bent under their bellies, where it enters ſo exactly into a cavity which is in the ſhell of the belly that it can hardly be diſtinguiſhed. That of the male diminiſhes in ſize quite to the end. The female's is equally large to its extremity. As faſt as the female lays her eggs, they attach themſelves to the long rough hairs with which the under part of the tail is furniſhed. Theſe ſupport, cover, and prevent them from falling, or from being detached by the ſand, herbs, or other unequal ſurfaces which fall in her way.

Theſe animals were ſo numerous in this marſhy ground, that it was impoſſible to ſet down one's foot without cruſhing ſeveral of them. I cannot ſay whether the inhabitants of the coaſts eat theſe animals, as they do in the Antilles, where they are of great ſervice to the Carribbees and negroes. The Creoles themſelves, according to father Labat, regale themſelves with them.

About ten o'clock in the morning, on Sunday the eleventh of December, we received the Governor on board. He left the town by two in the morning, in his canoe, with his ſon, accompanied by a miniſter from the King of Portugal, who was firſt preſident of the ſovereign council of Rio Janeiro, by the Oviodore, the major, and ſome other officers of the garriſon. The tide and a contrary wind had prevented their arriving ſooner.

The

The tent was pitched upon the quarter-deck, which was ſet out in form for the company; the ſhip was ſpread with the quarter-cloths, and the French flag diſplayed. When every body was ſeated, I ſaid maſs with the uſual ſolemnities; and at noon a dinner was ſerved up as elegant as poſſible in our preſent ſituation. The King of Portugal's health was drunk, under a diſcharge of eleven pieces of cannon, which were all anſwered regularly from the fort of Santa Cruz.

After dinner ſome inferior officers, the ſame who had played on different inſtruments at the Governor's, played again, and made a kind of concert, which laſted about two hours. During this amuſement, a wind aroſe unfavourable to the Governor's return, the weather became overcaſt, and ſo heavy a rain fell that it was ſcarce poſſible to think of going back to the town, or even to one of the forts. M. de Bougainville propoſed a party of play to amuſe theſe gentlemen, which was accepted. The wind and rain continuing, M. de Bougainville perſuaded the governor and his company to lie on board. While we were waiting for ſupper, which conſiſted only of the remains of the dinner, and at which the Governor, as well as moſt of the reſt of the company, drank only a glaſs of water, I converſed all the time with the Portugueſe miniſter, who wanted nothing but practice to ſpeak the French language well, and who made uſe of very expreſſive Latin words, whenever he was at a loſs for a French term. This miniſter, who was well acquainted with Brazil, and was at that time upon his tour through the country, according to cuſtom, was ſo obliging as to anſwer all my queſtions, and gave me all the information I could wiſh relative to the country and its inhabitants, of which I ſhall give an account hereafter.

Meſſieurs de Bougainville, and de Nerville, gave up their beds to the governor, and the Portugueſe miniſter; the Oviodore lay in the cabbin of M. Duclos Guyot our captain, and the other officers would abſolutely lie upon the table under the tent, where matraſſes were ſpread for them. We all paſſed the night as well as we could. At four o'clock in the morning, the governor and his company ſet off in his boat for Fort Santa Cruz, to give orders that we might be ſupplied with oxen, &c. and every thing we wanted. At his going away, we ſaluted him with nine pieces of cannon, which were immediately returned by the fort.

Before we ſat down to ſupper, M. de Bougainville had begged the favour of the Governor, to get the letters we had written to give our friends ſome intelligence about us, conveyed to Portugal, and from thence to France. A few days after, a Portugueſe ſnow, which then lay at anchor before the town of *Our Lady in Exile*, intended to ſet ſail for Rio de Janeiro, and had promiſed us to deliver our packets to ſome one of the captains of the fleet, which was to ſet out from thence towards the end of the month. The Governor very readily took this commiſſion upon himſelf; and M. de Bougainville ſent him our parcels. But whether the Rio de Janeiro fleet was gone, before our letters reached that place, or whether from any other accident, it is certain that our parcels never came to hand.

Five or ſix hours after the Governor's landing at Fort Santa Cruz, a preſent came to us from him of two oxen, as many cows, a heifer, two turkeys, twenty-ſix Brazil ducks and drakes with large red tufts, and other refreſhments. M. de Bougainville had in the evening preſented him with a box full of ſnuff-boxes, painted and varniſhed by Martin, and with ſome very handſome fans.

The two following days were employed in compleating our proviſion of wood for firing; it was compoſed of ſaſſafras, cedar, and of yellow wood of Brazil. Having been told by a free negro, that the tree which bears the balſam of *copaiba*, known under the name of copahu, is not rare in this country, I did my utmoſt to get ſome of it, but could not ſucceed. The Portugueſe who had promiſed to procure me ſome, aſſured me that this balſam only flows during the full moon.

As our deſtination was for a country, where our ſailors had never yet been, and whoſe ſeas and weather were reputed tem-

peſtuous,

peſtuous, before we ſet ſail from Saint Catherine's, our captain, with his uſual prudence and foreſight, took care to provide himſelf with ſmall top-maſts and top-gallant ſails, to ſerve in caſe of bad weather.

For this purpoſe he applied to the free negro, of whom I have made mention more than once; he rendered us all the ſervice in his power with the greateſt readineſs, and even made a propoſal to M. de Bougainville of going with him; adding, that he was ready to go to any place where we pleaſed to take him. He was a ſtout fellow, and worked hard; M. de Bougainville would willingly have acquieſced in his demand, if he had not been apprehenſive that the Portugueſe might have complained we had put in to decoy away the negroes of the country; and that ſuch a report or ſuſpicion might be of diſſervice to thoſe French veſſels which might afterwards put in at St. Catherine's.

This negro went himſelf into the foreſt, to look out for ſuch trees as he thought might beſt ſuit our captain's purpoſe. When he had found them he informed him of it, and conducted us there through thickets and buſhes, in the midſt of which we were obliged to climb up the mountain to get at theſe trees, which were in the thick of the foreſt. We went thither well armed, and in a pretty large body, as well for the convenience of cutting down theſe trees, as to be able to convey them to the ſea ſide. We met with no wild beaſts in the foreſt, but only two or three large ſerpents which we killed. As we were going along I cut ſeven or eight bamboo canes; they are a kind of knotty reed, the joints of which are very cloſe, and the cloſer they are the more beautiful are the canes. Theſe bamboos were of a good ſize, from five feet and a half to ſix feet long; but unfortunately were not come to their maturity. In drying them they ſhrivelled up, and became as it were fluted. I left them at St. Malo's.

In order to convey our wood to the water ſide, we were obliged to tie them with cords, and to drag them, ſometimes even to lift them, over branches which ſtopped up our way. We were often forced to make a road, which we did with hatchets. I obſerved palm trees there of a prodigious heighth, of about a foot diameter, and as ſtraight as bulruſhes; this is a kind of tree whoſe trunk and branches are quite covered with ſmall thorny excreſſences, of about ſix or eight lines in diameter, or ſometimes more at their baſe, projecting about half an inch, and the thorn, which grows in the center, being about four lines long. The bark of this tree is grey, and reſembles that of the beach tree. May not this be the ſame, which is in our Antilles called *bois epineux?*

On Tueſday morning a ſailor, after having cut ſome graſs for our cattle, ſitting down by it with his legs bare, was bit near the ancle by a ſerpent, as he told us, about a foot and a half long, of a reddiſh yellow colour in ſtreaks. He paid no attention to the bite, and as ſoon as he came on board he dined heartily without uneaſineſs, ſaying he was hungry. In about half an hour after he found himſelf ſick; and perceiving his leg very much ſwelled and painful, he came and acquainted me with it. I began by endeavouring to keep up his ſpirits, and to eaſe him of the fear which had ſeized upon him. Whilſt I was giving notice of this to the two ſurgeons of our frigate, he vomited, and did the ſame once or twice in the ſpace of an hour. We made him take two drams of theriaca mixed in a glaſs of wine, with ten drops of volatile ſpirit of ſal ammoniac, and after having ſcarified the wound, which was already become black, applied to it a plaiſter of theriaca pounded with garlick. Notwithſtanding this his ſickneſs continued, and he vomited two or three times more. The ſame remedy was repeated. In the meanwhile, a Portugueſe officer from Fort Santa Cruz came on board, to whom we related what had happened. The ſailor's account, with the deſcription of the reptile, gave the officer reaſon to judge, that this ſerpent was of one of thoſe kinds which the people of the country call *Jararaca*. "Its venom, ſaid he, is ſo dangerous, that it cauſes inevitable death to thoſe, who are not excited by it to vomit within the four and twenty hours. But ſince

ſince your ſailor has vomited, you may make yourſelves eaſy upon his account. Continue however to give him the ſame medicine, with the addition of an emetic. There are ſeveral other kinds of *Jararaca*, which are equally to be dreaded; eſpecially one, which is of an earth or cinder colour, with ſome ſtreaks on the head which are rather darker." The next day, neither the blackneſs of the wound, nor the inflammation of the leg being increaſed, the emetic was given to the patient, and a freſh plaiſter of the ſame kind as the former was applied. After this no other accident happened to him, and the wound was treated as a common one. He was purged twice afterwards, and from that time was very well. In going into the woods and fields, you are almoſt always liable to be bit by theſe dangerous reptiles, which are very numerous there. We very frequently ſaw in the ſand on the ſea ſide, winding furrows formed by the traces of ſerpents which had paſſed there. If any one who has the miſfortune to be bit by one of them, does not immediately meet with proper aſſiſtance, he muſt expect to die in the moſt cruel tortures. Some ſorts, eſpecially thoſe of the Juraracas, exhale a very ſtrong ſmell of muſk. This ſmell is of great ſervice to thoſe who know it, to prevent their being ſurpriſed by them.

The only lizard which I ſaw in the iſle of St. Catherine's, might be about two feet long, and three or four inches broad. Its ſkin was black, ſpotted with white from the head to the end of the tail. The belly was much the ſame, but the white was rather prevalent; all over the reſt of the body, the black and white was almoſt equally diſpoſed in ſpots of a regular figure: its ſhape in other reſpects was like that of the green lizards in France. M. de Nerville, who was with me, was preparing his gun to fire at him, when I perceived that the animal was dead. We went towards it; but as it already ſtank very much, we did not think proper to examine it with greater attention. Might not this be what is called by the people of the country the *Maboya*, or *Tejuguacu*, and *Iguana* by Piſon and Margraff?

The

The rattle ſnake is very common there: its length extends as far as three feet, but very ſeldom exceeds above half a foot more. Its colour is of an iron grey, regularly ſtreaked. At the extremity of its tail is faſtened what the Spaniards call its *caſcabelle*, and for the ſame reaſon the ſerpent *caſcabella*. This caſcabelle, which we have choſe to call the rattle, on account of the noiſe it makes, reſembles the huſks of peas dried upon the plant. It is divided in the ſame manner into ſeveral joints, which contain ſmall round little bones, whoſe friction produces a ſound much like that of two or three rattles, or ſmall bells which make but a dull noiſe. The day that M. de Belcourt and I were in the woods looking for the animal whoſe traces we had ſeen in the ſand, we thought we heard this ſound mixed with a kind of hiſſing; which alſo was much like the noiſe uſually made by graſhoppers. The bite of this ſerpent is ſo dangerous, that it is happy for the inhabitants of the countries where it is found, that nature has given to this reptile a ſign to warn them of its approach; without which, its colour differing very little from that of the earth, it would be very difficult for them not to be ſurpriſed by it, and to avoid it. This animal is alſo called *boicininga*.

To feed the cattle which we were taking with us from the iſland of St. Catherine, we laid in a proviſion of the ſtalks of the *banana tree*, with which we covered our quarter-deck, both within and without. This forrage is the moſt convenient for tranſportation; as it takes up very little room, and was therefore leſs cumberſome than any other. Beſides, meadows are very ſcarce in this iſland, as well as along the coaſt of the Continent, and the little graſs which grows there is marſhy. It would have been difficult to have procured a ſufficient quantity of it; and that even, not being come to a ſtate of maturity fit for keeping, would have heated, and afforded a very bad ſubſiſtence for theſe cattle. The ſtalks of the banana were a very good ſubſtitute, as much on account of their keeping very well, as becauſe they are very nouriſhing. We had only to cut them in pieces with a knife, a knife, after which our oxen and ſheep ate them with great eagerneſs.

The banana is a plant, the ſtem of which is compoſed only of leaves rolled one upon the other, of a reddiſh white colour in ſome places, and a greeniſh yellow in others. When the root ſhoots out a ſprig, there are only two leaves rolled one within the other, which appear coming out at the edge of the ground. Theſe leaves unfold and expand themſelves, to give way to two others, ariſing from the ſame center. Theſe being rolled up as the former, expand themſelves in the ſame manner, and are followed by ſeveral others, which growing higher and broader as they ſucceed each other, and being always rolled up in the ſame manner, form at length the ſtem of this arboreous plant, which riſes to eight, ten, and even twelve feet; but not any higher. Then the leaves grow out from the top, and the middle of the ſtem, to which they are connected merely by a foot ſtalk about an inch in diameter, a foot long, round on one ſide, and hollowed on the other by a groove in the middle. This foot ſtalk being continued, forms the vein or band which runs along the middle of the leaf, which is ſometimes fifteen or eighteen inches wide, and ſix or ſeven feet long. The ſupine diſk of this leaf is of a fine green colour, the prone diſk of a green inclining to grey, which makes it appear ſilvered. It is nearly of the ſubſtance of very thick parchment; yet it is ſo delicate, and its ſize expoſes it ſo much to the action of the wind, that it is divided into ſeveral ſlips. Theſe ſlips extend from the vein running in the middle towards the edges of the leaf, by the ſide of the ſmaller veins running in the ſame direction, and appearing, ſome like narrow ſilver ribbands, others like ſlips of the ſame colour, fixed to the vein in the middle, and rolled upon themſelves.

When the banana is grown up to its natural height, it is from nine to ten inches in diameter, and the ſtem of it is ſo tender, that, though the leaves of which it is compoſed, are joined very cloſe to each other, it may eaſily be cut with a knife, or even with a ſingle ſtroke of a hedging bill, taken a little aſlant; for

3 the

the leaves are pulpy and full of juice, and this is the reaſon why it always thrives beſt in a rich and moiſt ſoil.

When it is come to ſuch maturity as to be capable of bearing fruit, it puſhes out, from the middle of the top of its ſtem, another ſtem of about an inch and half in diameter, and three or four feet in length, which is covered with circular rows of buds of a yellow colour bordering on green. A large bud in the ſhape of a heart, from ſix to ſeven inches in length and three in diameter, terminates this ſtem. It is compoſed of ſeveral pellicles laid one over another, the outſide of which is red, and has a ſecond covering which is compact, ſmooth, and of the colour of the lilack. This bud riſes from an aperture made by the diviſion of the ſtem into four parts. At firſt the ſtem is ſtrait, but in proportion as the bloſſoms diſappear, and give way to the fruit, which ſucceeds them, the increaſe of weight bends it inſenſibly, and draws it more and more towards the ground.

In our Antilles, this ſtalk laden with fruit is called a *régime.* I know not by what name the Portugueſe call it. One of theſe ſtalks is ſometimes furniſhed with as many bananas, as one man can carry. The fruit is faſtened to the part which before ſupported the flower. It is cuſtomary to cut off the ſtalk, as ſoon as the fruit upon it begins to change from green to yellow. It is then ſuſpended in an airy part of the houſe, and the fruit is eaten as it grows ripe, which is diſcovered, by its giving way to the finger, and becoming yellow. We hung up about a ſcore of theſe ſtalks round the quarter deck; and ſome of our officers were ſo fond of this fruit, and eat it ſo eagerly, that they would not give it time to ripen. The banana is about two inches in diameter, and the longeſt I have ſeen of them did not exceed ſix inches in length. The two ends terminate in a rounded point: the figure of it is angular, but the angles are very obtuſe. The ſkin is ſmooth, pliable, rather thicker than that of a fig, and much more firm. The pulp is of a yellowiſh white, and of the conſiſtence of very fat new cheeſe, blended with its cream; or of butter recently churned, which the banana reſembles very much,

eſpecially

eſpecially when roaſted. It taſtes indeed much like this kind of butter, ſuppoſing it to have been mixed with the pulp of quinces rather too ripe. It is eſteemed an excellent kind of food; for my part, I found nothing extraordinary in it; though I eat the fruit both raw and roaſted, ripe and unripe, in order to judge of the difference of the taſte.

We had been very deſirous of providing ourſelves with a quantity of *batatas* and *yams*, but they were not yet in a ſtate fit to be taken out of the ground. The *batata* is a ſpecies of potatoe, or topinambou, but much more delicate.

The *yam* is a creeping plant, furniſhed with branches which take root, and ſpring up again without cultivation; ſo that if care is not taken to root out a number of them, they will ſoon ſpread all over the ſoil, though there ſhould not have been more than one or two roots planted at firſt. The ſtem is ſquare, of the ſize of one's little finger, or thereabouts. Its leaves are cordiform, having their apex a little lengthened out and pointed. They are of a dark green colour, as large as thoſe of the *lappa major* or greater bardana. They grow leſs in ſize as they are placed farther from the root; but they are ſtill ſmooth, thick, and pulpy, fixed to the ſtem in pairs, by ſhort petioles, ſquare, and rather curvated. From the ſtem ariſe ſome cluſters of ſmall campanulate flowers, the piſtil of which becomes a ſiliqua or pod, filled with ſmall black ſeeds. Theſe ſeeds are ſeldom ſown, becauſe the plant thrives better and faſter from ſlips. For this purpoſe, the head of the fruit, with part of the ſtem ſupporting it, is put into the ground.

The root is more or leſs thick in proportion to the goodneſs of the ſoil in which it grows. The rind is unequal, rough, thick, of a deep violet colour, and very hairy. The inſide is of the conſiſtence of beet-root, of a greyiſh white, bordering ſometimes upon a fleſh colour. It is eaten prepared in the ſame manner as the beet-root, boiled in water, or roaſted on the embers, ſometimes with the meat. It is well taſted, very nouriſhing, and eaſy of digeſtion. The negroes and Portugueſe are remarkably fond of it.

One of the two married Acadians, who were lodged in a hut on ſhore, was afflicted with a violent diarrhœa, which baffled all the ſkill of the two ſurgeons of our frigate. The freed negro offered to cure him with a ptiſan, and might probably have ſucceeded, if he had had more time for the application, for the man found himſelf much better after having taken it only for two days. What this negro called a ptiſan, was no more than a ſimple decoction of the ends of the buds and infant fruit of the guaiava. If this fruit had been a little farther advanced, perhaps it might have been ſtill more ſerviceable. The Acadian not having laid in a ſtock of theſe buds, before we ſailed, had it not in his power to continue the medicine: his diſorder returned upon him with greater force, and did not leave him till about a fortnight before our departure from the Malouine iſlands, where he began to find himſelf better a few days after our landing. The wholeſomeneſs of the air, added to the exerciſe he took, made him ſtronger every day, and at the time we ſet ſail from thence, he thought himſelf perfectly cured.

The ſame negro had cured the Acadian's wife's ſiſter, whoſe name is *Benoit*, in a few days of an inflammation in her legs, which had got to ſuch a height, that ſhe could ſcarcely ſtand. This inflammation was attributed to a ſcorbutic habit. However this was, ſhe complained of great pains in her ancle-bones, which went off after the negro had applied a fomentation made of ſome herbs of the country boiled in clear water. In ſix or ſeven days ſhe was cured; and they aſſured me, that the negro had uſed nothing but the guaiava.

The guaiava is a tree well known in our American iſlands. Thoſe which were called by that name on the iſland of St. Catherine's, were not more than eight feet high, and the trunk between ſeven and eight inches in diameter. None of thoſe I ſaw were of a larger ſize. The bark of it was ſomething whiter than that of he apple tree, its branches extended in the ſame manner,

manner, and the fruit, which was very young, reſembled apples of about a month's growth. By the leaves and the ſhape of the tree, I took it at firſt for a quince-tree. I was told, that the fruit when ripe is excellent, and it was repreſented to me as of the ſame kind as our guaiavas of the Antilles; although the deſcription, which father Labat gives of this tree and its leaves, does not anſwer to the guaiava tree of Brazil. In other reſpects, the Portugueſe aſcribe the ſame properties to their guaiava tree, as father Labat does to that of Martinico.

The ſame Acadian, when he came on board again found himſelf very much incommoded with a ſmall tumor, which had come a few days before on the great toe of his left foot. This tumor increaſed, and the pain it occaſioned, increaſed in proportion. He ſhewed it to the doctor (for ſo we call the two principal ſurgeons on board) who immediately diſcovered it to be *anigua* or *nigue*, or the *pique*, as it is called at Peru. This is an inſect ſo exceedingly ſmall, that it is ſcarcely viſible. A particular deſcription of it may be found in M. Ulloa's voyage to Peru, which exactly agrees with what we ſaw at the iſland of St. Catherines. Our Acadian was cured by extracting the neſt, and applying tobacco aſhes to the part. The legs of this inſect, ſays the author I have juſt referred to, are not formed for ſpringing like thoſe of fleas, which is a very providential circumſtance, for if it had the power of leaping, there is no living creature in the parts where theſe inſects are found, but what would be full of them. Such a breed would deſtroy three fourths of mankind, by the different accidents it might bring upon them.

The nigua always harbours in the duſt, and particularly in dirty places. It faſtens upon one's feet, even upon the ſoles of them, and upon one's fingers, and pierces the skin ſo ſubtilely, that it makes its way almoſt without being felt. It is ſeldom perceived, till it begins to extend itſelf. At firſt, there is no great difficulty in pulling it out; but if it has once got in only its head, it fixes itſelf ſo firmly, that it cannot be got rid of without the loſs of ſome of the parts adjacent. If it happens not to be diſ-

 covered

covered ſoon enough, it pierces through the firſt ſkin without reſiſtance, and lodges itſelf there, where it continues, ſucking the blood, and makes itſelf a neſt of a fine white membrane reſembling in ſhape a flat pearl. Within this ſpace it lies ſo that its head and feet are turned towards the outſide for the convenience of nouriſhment, and the reſt of its body towards the inſide of the membrane in order to depoſit its eggs there. As the eggs are laid, the membrane grows larger; and in four or five days time, becomes one ſixth of an inch in diameter. It is of the utmoſt conſequence to remove this, otherwiſe, it will burſt, and ſpread about an infinite number of ſpawns, like nits; that is to ſay, ſo many niguas, which will inſinuate themſelves preſently into the parts about, and create a great deal of pain, not to mention the difficulty of diſlodging them. Sometimes they will penetrate even to the bone; and after one has ſucceeded ſo far as to get rid of them, the pain continues till the fleſh and ſkin are entirely healed.

The operation is tedious and painful. It conſiſts in ſeparating with the point of a needle the fleſh which touches the membrane, wherein the eggs are contained; and this is not eaſy to be done without cracking the membrane, a circumſtance abſolutely neceſſary to be guarded againſt. After having detached every ligament even to the ſmalleſt, the pearl is next to be removed, which is larger or ſmaller in proportion to the time the inſect has been lodged there. If unfortunately the neſt ſhould be broken, double care muſt be taken in ſeparating all the roots, and particularly in ſecuring the principal nigua; who, if ſhe eſcapes, will begin again to lay her eggs before the wound is cloſed, and burying herſelf in the fleſh would make it much more difficult to remove her. In the cavity made by the tumor, they put ſome hot aſhes of chewed tobacco.

Although this inſect is not felt at the time of its penetrating through the ſkin, by the next day it cauſes a violent and very painful itching, particularly in ſome parts, ſuch as under the nails.

The

The pain is not ſo great at the bottom of the ſoot, the ſkin there being thicker.

It is obſerved that the nigua has a mortal antipathy to certain animals, particularly the *Cerda*, which it devours by degrees, and whoſe fore and hind feet are found after its death full of holes.

Notwithſtanding the ſmallneſs of this inſect, it has been obſerved to be diſtinguiſhed into two ſpecies, the one venomous, the other not. The latter is of the colour of fleas, and makes the membrane white, in which it depoſits its eggs. The other ſort is yellowiſh, and its neſt is of the colour of aſhes One of the properties of this ſpecies is, when it has inſinuated itſelf into the extremity of the great toe, to cauſe a very high inflammation in the glands of the groin, accompanied with ſharp pains, which do not go off till the eggs are extirpated. It is to M. de Juſſieu that we owe this diſtinction of the niguas into two ſpecies. That gentlemen, as well as the other French Academicians, who accompanied him in his voyage to Peru, having had the mortification of repeatedly experiencing theſe pains, which they knew not how to account for.

During the great heats, the utmoſt care is neceſſary to avoid wetting one's feet. Without this precaution as appears by experience a man is liable to a diſorder of ſo dangerous a nature, that it is generally mortal.

On Wedneſday, the 14th of December, having got our proviſions on board, and the wind being ſoutherly, at ten in the morning we unmoored and ſent our longboat on ſhore for the reſt of our things, and the Acadians. Before they were taken into the boat, their baggage was examined with great care to ſee if no *Cancrelas* had got in amongſt it, ſome of them having been ſeen in their hut. Theſe are inſects of the ſize of a May bug, and ſomething of the ſame form, though rather flatter and longer, having a coat of a very dark green, but neither ſo hard nor ſo ſolid. They do an incredible deal of miſchief in ſhips, as they multiply very faſt, and lodge themſelves every where,

where, eating through paper, books, cloaths, bifcuits and even wood itfelf. They fpoil every thing with their excrements and the difagreeablenefs of their fmell. As fome of them had been feen in the hut, where the Acadians lived, it had been recommended to them to clean their cloaths thoroughly: and we had the good fortune to be as free from them as from other infects ufually met with at fea. At the Antilles, thefe infects are known by the name of *Ravets.*

At the fame time we difpatched M. Alexander Guyot to Fort Santa Cruz, to make our apology to the Commandant for not faluting him at our departure, which we could not do on account of the live ftock we had on board.

At eleven we fet fail, and after doubling the point of *Bon-port* anchored in fix fathom water, muddy ground, about two thirds of the diftance of one fhore from the other towards the north fide; here we waited for our longboat and yawl, and to take in ten oxen, which was finifhed by five in the afternoon. We lay to after this with one anchor all night, the weather being dull and hazy.

Thus we took leave of the ifland of St. Catherine's, where, as has been feen, we did not meet with the fame caufe of complaint as admiral Anfon had done. It might be made an excellent habitation, if they would take the pains to clear it. Befides the little town I have fpoken of, there are only a few huts or plantations on the coaft of the ifland and continent adjacent. All the reft is a foreft covered with lofty trees, and, as it were, choaked up with underwood. On the ifland particularly, the fpecies of thorny aloe grows fo thick, as to render it in many parts almoft impenetrable. On the fkirts of the woods towards the coaft, there are many different forts of trees of the height and thicknefs of apple-trees, but whofe leaves are for the moft part fmooth, of a beautiful green, and fhaped like thofe of the wild laurel. There is fcarce any other difference between them than in their height or thicknefs. I obferved one amongft them, which at firft fight might have been taken for an almond

tree:

tree: but the leaves were a little too large: the fruit had the appearance of a green almond, but on a cloſe examination was found to be made ſomewhat in the ſhape of a heart. Another bore a flower or fruit, reſembling the veſicles or membranes which encompaſs the fruit of the plant called *Alkakengas*. This Brazilian fruit is of the ſize of a ſmall nut, and the outſide is of a whitiſh yellow. It opens itſelf into four parts, and is compoſed of ſeveral ſimilar coats, whoſe internal ſurface is of the colour of the fineſt carmine. Theſe barks, or parts of the fruit or flower are ſo ranged, that the middle of the upper coat forms a covering to the edges, by which the four internal diviſions are connected. Theſe coats are eight in number, four external, and four internal, each of them about the thickneſs of the twentieth part of an inch: on the inſide is faſtened to the center, a ſmall white ball, which is undoubtedly the piſtillum; if it is a flower I am deſcribing. I ſhewed it to a Portugueſe, who could not inform me either of its name or properties.

The *caraguata* is a plant very common in theſe woods, and moſt of the rocks upon the coaſt are covered with it. It is likewiſe found in great plenty upon the branches of large trees, like the miſletoe upon our apple-trees and oaks. It has a long, ſharp, prickly leaf, almoſt like that of the flags, which plant it reſembles both in the ſhape of its leaves and their ſituation, as they all come out from the root: but this produces a round ſtem furniſhed with ſome leaves that are of the colour of the fineſt carnation, as is likewiſe the tip of ſome of the internal leaves of the tuft, which are neareſt to the ſtem. At the top of this ſtem are produced ſpikes of flowers of a lively red, which are ſucceeded by a kind of fruit half an inch long, of the thickneſs of a large quill, and of a violet colour. It contains a white viſcous ſubſtance full of ſeeds, which are flattiſh, reddiſh, and very ſmall. I ſhould imagine, that the caraguata is a ſpecies of the algæ or flags. In a narrow path traced out near the border of the woods, we found ſome plants called by the Brazilians *juquiri*, and *caaco*, and by us *ſenſitive*. Of theſe we ſaw two ſorts; that which I am going to ſpeak of, produces ſtems about two feet high, full of branches, and ſtriated, nearly quadrangular, of a green colour, and pretty well defended with ſmall yellowiſh prickles. The leaves are ſtationed oppoſitely upon the pliant branches: the prone diſk is of a whitiſh green, the ſupine diſk of a pale green. From the length of the ſtem ſeveral branches puſh out, whoſe extremities are ornamented with a ſmall round head, which is hairy, and of a whitiſh purple. This is the flower, which is ſucceeded by a huſk or ſmall pod, crooked, and of a cheſnut colour; when ripe, it is covered with ſmall white hairs.

The ſecond ſort does not riſe to any great height from the ground; I have only ſeen it in the ſandy lands along the coaſt: it ſeems to creep as it were, and does not differ from the former. Both of them ſhut their leaves and let their branches fall, as if they were blaſted, almoſt at the inſtant they are touched with the hand. The next moment they appear again in their full vigour. It is from this circumſtance, no doubt, that they have obtained the names of *the chaſte herb*, *herba caſta*, *mimoſa*. The leaves when eaten are a deadly poiſon, which can only be counteracted by eating the root itſelf. The ſame leaves, when applied as a cataplaſm, are a cure for ſcrophulous humours.

The ſoil of St. Catherine's iſland, and the coaſts of the continent are ſo prodigiouſly fertile, that the beſt fruit grows there in abundance almoſt without culture. The foreſts abound with odoriferous ſhrubs. The road affords a very great variety of excellent fiſh. The figures of thoſe, to which our ſeamen have given the names of *Balaou*, *Lune*, *Brune*, *Lame d'épée*, *Crapaux de mer*, &c. may be ſeen in the plates. The long beak by which the *Balacu* is diſtinguiſhed induced me to name it the *Beccaſſine de mer*, or ſnipe-fiſh. The extremity of this beak which is very ſolid, and as hard as that of a bird, is about a quarter of an inch in length, and of the colour of the fineſt vermillion. Its body is almoſt tranſparent: a fillet or ſtripe of greeniſh blue runs from the gill, to the tail: its ſcales are ſo fine that they are hardly

 diſcernable.

difcernable. Its flefh is firm, and has an excellent flavour. This Lune, or moon-fifh, appears to be covered with a fheet of filver.

The *Lame d'épée*, or fword blade, could hardly have a name better adapted to its figure. The *Crapauv de mer*, or porcupine fifh, might have been called fea-urchins their body being covered with prickles about two lines in length. In the fpecies which approached the neareft to the ordinary fhape of fifhes, the jaw was armed with teeth, which were pretty large and flat like the canine-teeth in the human fpecies, and was not unlike the human mouth even in the lips.

We did not catch any beautiful fhell-fifh here; the only one deferving notice was a helmet fhell, which was at leaft eight inches in diameter. We met with a foldier-fifh, and fome fmall fea-horfes. Our fifhing was always accompanied with fear; on account of the fharks which very much infeft this road.

The fharks taken by us, a fpecimen of which is exhibited in the plates, were not of an extraordinary fize; they were of that fpecies called the dog-fifh. On an attentive examination of their rows of teeth, we thought they amounted to feven in number, inftead of fix, which are generally attributed to them. They were flat, triangular, fharp, and their edges were ferrated. They did not appear to be firmly fixed in the jaw like thofe of other animals. They were moveable, opening and fhutting like the fingers, in fuch a manner that each row in recovering its fituation lay over the next to it, fo that the upper row bending towards the inner part of the gullet, filled up the vacuity or interval between the lower row of teeth. They are difpofed like the flates of a roof, or, perhaps, like the leaves of an artichoke.

It is faid that the fhark is conftantly preceded by another fifh called the *Pilot*: we can affirm the contrary; at leaft we have feveral times feen fharks without this harbinger.

The pilot is among the number of beautiful and good fea fifh It is of a blue colour difpofed in ftripes; fome of them to the number of fix are of a fine blue, which upon the back is of a deep caft, but becomes gradually lighter as they verge towards the belly. Four other ſtripes, together with the head and tail, are of a very clear blue, or of a white daſhed with blue. The eyeball is of a fine gold colour, except the pupil which is black; the two forks of the tail are white.

I did not recollect this fiſh by the deſcription given of it in M. Valmont de Bomare's dictionary of natural hiſtory. It ſeems rather to belong to ſome ſpecies of the *Remora*. Is he miſtaken in this article, as he is in ſeveral others? Or have I been impoſed upon by being told that a fiſh was called the *Pilot*, which has no title to that name? See the figure of it in the plate.

As we did not ſee any ſharks preceded by pilots; ſo neither did we catch any one that had not upon it ſeveral ſucking fiſhes faſtened cloſe to it about the head. The Brazilians call the ſucking fiſh *Iperuquiba*, and *Piraquiba*, the Portugueſe, *Piexepogador*. The largeſt that we caught was about eight inches long, and two and a nalf over in the broadeſt part. The upper part of the head, which is two inches long, is flat, reſembling an ox's palate, furrowed acroſs, and faſtened to it in ſuch a manner that the edges ſhould not adhere. Theſe furrows are armed with prickles ſo hard and ſolid, that when rubbed upon wood, they act as a fine file. By means of theſe, the ſucking fiſh faſtens himſelf ſo ſtrongly about the gills and belly of the ſhark, that he ſuffers himſelf to be taken along with him. He cannot even be ſeparated without a knife or ſome other inſtrument. The under jaw is longer than the upper. This fiſh has ſmall eyes of a gold-coloured yellow, and the pupil is black. An infinite number of ſmall tubercles, that are pretty ſolid, ſupply the place of teeth. Near each of the gills is a triangular fin about an inch long: there are two others near the belly, which unite at the place of inſertion, and one under the belly and another upon the back, which extend from the middle of the body to the tail. Its ſkin is ſmooth and ſlippery like that of an eel, and of the colour of brown ſlate.

Many have miſtaken the back of the ſucking fiſh for the belly, on account of the part by which it faſtens itſelf to the ſhark.

ſhark. I have obſerved it with all the attention I was maſter of, and am convinced of the miſtakes of authors as to this particular; as appears by the plates.

One of our officers had a preſent made him of two young *Tucans*, called by ſome *Tulcan*, and by the Spaniards in the iſthmus of America, *Preacher*, becauſe this bird, it is ſaid, perching on the top of a tree while others are aſleep below, makes a noiſe with his tongue reſembling words ill articulated, and ſpreads the ſound on every ſide, leſt the birds of prey ſhould ſeize the opportunity of devouring the others while they are aſleep.

The Toucan is pretty nearly of the ſize of a wood-pigeon; but ſtands higher on its legs, which are of a blueiſh grey as well as its feet, which are armed with pretty long claws; its tail is about four inches long, ſometimes black, and rounded at the end; but uſually variegated with blue, purple, and yellow upon a dark brown. The back and wings are of this laſt colour, except ſome black feathers in the wings. Its head, though very large, is very ſmall in proportion to its bill, which is between ſeven and eight inches long from the place of inſertion to the end. The upper part near the head is about two inches at the baſe, and as it lengthens forms a figure pretty nearly triangular, and at the ſame time convex at the upper part, the two lateral ſurfaces being a little raiſed and rounded. The upper one which forms the inſide of the bill is hollow, having ſerrated edges or lips. The lower one is ſhaped like the upper, only it is rather concave underneath. Theſe two are of an equal length, are inſerted into each other, and grow gradually leſs towards the extremity, which is rather crooked and ſharp underneath. The tongue is a whitiſh membrane almoſt as long as the bill, but very narrow and flat, and has the appearance of the point of a pen; its eyes are round, beautiful, lively, and ſparkling, and are inſerted into two bare cheeks, which are covered with a ſky-coloured membrane. In ſome the iris is of a clear blue encircled with white, in others it is quite black. There are different ſpecies of them, at leaſt they differ from each other in the colour of their bill, and their feathers. The bill of ſome is green, having a black circle and two white ſpots near the root. That of others is black, the inſide red, with a greeniſh yellow circle near the head. They are very common in this country. We were likewiſe aſſured that great numbers of pheaſants were found here; but we ſaw none. The green parrots fly here in flocks like the ſparrows in France. We killed great numbers of them, and found them to be as good as the pigeons in our dove-cotes. Lions, panthers, leopards, ounces, and tygers, infeſt the woods and make travelling dangerous. It is happily very ſeldom that they come near inhabited places. The water of the rivers is of an excellent quality. But all theſe advantages are defeated by the inconvenience ariſing from a very unwholeſome air, which is probably the cauſe of the pale complexion of the white people who inhabit this country. From theſe woods, which the ſun never penetrates, groſs vapours ariſe without intermiſſion, which form continually thick fogs on the tops of the mountains ſurrounding the iſland. The low grounds which are very marſhy, are equally miſty from between ſix and ſeven in the evening, till the ſun diſperſes the vapours at eight the next morning. Theſe vapours frequently ſmell like mud, and as there is no free circulation of air, they ſeem to diſperſe only to make room for others which ſucceed. This unwholeſome atmoſphere is undoubtedly corrected, at leaſt in ſome meaſure, by the multitude of aromatic plants, whoſe ſweet ſmell is perceived three or four leagues at ſea when the wind ſets off the ſhore. Our dogs informed us of our approach to land at this diſtance at leaſt, by ſmelling towards that ſide for near half an hour. We were likewiſe regaled with the perfume. It is obſervable, that dogs are of great uſe on board a ſhip in diſcovering its approach to land. Ours never failed to go upon the fore caſtle, where they turned their noſes to ſmell on that ſide neareſt the land, though at the diſtance ſometimes of between five or ſix leagues, when the wind ſet towards us from the ſhore. They would even give us notice of a ſail, at the ſame diſtance, if to windward of us. They ſtayed there about half a quarter of an hour, and returned more than once to the place after they had quitted it.

Beſides the birds already mentioned, we frequently met with *Criards*. The Portugueſe give this name to a kind of crow, whoſe plumage is of a fine pale blue. Theſe, they ſay, are the ravens of the country. Their ſhape is the ſame, and they are almoſt as unpalatable. The *Tiépcrangas* are of the ſize of the thruſh; their wings, tail, and part of their bill are of a deep brown; the reſt of their plumage is of a beautiful vermilion tinctured with carmine, and inclining to ſcarlet. They are called by ſome of our ſeamen Lorys, by others *Cardinals*; but this laſt name belongs to a bird of nearly the ſame ſhape, whoſe plumage is intirely red without any mixture of brown.

On the morning of our departure from St. Catherine's as we caſt anchor when it was almoſt dark, we perceived all along the ſhrouds, haliards, and the reſt of the tackling, a number of ſmall moving lights, or rather ſmall moveable lamp-glaſſes. Though we concluded that they were fire flies, our ſurprize was the greater as we had not ſeen any of them before. They had four wings, two of which were tranſparent like thoſe of our common flies, and two opaque, ſmooth, brown and ſolid like the outer ones of the May-bug, and like thoſe, ſerving as a caſe to incloſe the under ones. The head is black, ſhaped like the trefoil, and furniſhed with two antennæ which are likewiſe black, four lines long, and ſeem to be compoſed of ſmall horns inſerted into each other by their points. The eyes which are placed near theſe antennæ, are round, black and firm as horn; they are ſparkling and prominent, and of the ſize of the ſmalleſt poppy-ſeed. The body and legs which are ſix in number, are of a darkiſh brown. With the naked eye one may eaſily perceive ſix rings gradually decreaſing in ſize from the neck to the extremity of the body, which terminates in a rounded point. Theſe rings are as ſolid as thoſe of which the body of the May-fly is compoſed. The largeſt ring which forms all the fore part of the

body

body, to which the legs are articulated, is ſomething more than two lines broad, and two in length, and is covered with a down or light duſt like the wings of butterflies. From this part, and from the head proceed rays of light, reſembling that of the glow-worms which are ſeen in France during the ſummer ſeaſon, and at the beginning of autumn.

I put one of theſe flies into a piece of paper when I went to bed, intending the next day to make a drawing of it. But when I was going to ſet about it, the fly was not to be found. It had gnawed its way through the paper, and made its eſcape. The day following, when I was laid down, I perceived a light in one of the places where I had put my books. I thought no more of the fire-flies, and imagined at firſt that this light proceeded from the lamp in the binnacle, which was near the window of my cabbin; but perceiving the light change its place, I recollected that it muſt be the fly which had got away the preceding night. Having caught and incloſed it in a glaſs veſſel, the next day I obſerved it at my leiſure, and drew a sketch of it.

When we left Montevideo, we ſaw the ſame appearance upon the river of Plata, or *Rio de la Plata*, during the unexpected calm that happened on the day of our departure. As theſe diffuſed, a more brilliant and ſparkling light than thoſe of the iſland of St. Catherine, I put ſome of them with freſh graſs into a glaſs veſſel which I incloſed in another, and having placed it upon my table, fetched a book which I read with great eaſe without the aſſiſtance of any other light, though the print was very ſmall.

The next morning I took one out of the veſſel, and transfixing it with a pin, which I ſtuck into the table, I made a drawing of it.

It was four lines broad, and eleven and an half long including the cap of three lines which covered the head.

The body was furniſhed with four wings. The two upper ones were of a fine black velvet lace, having a ſtreak of gold coloured yellow near the outward edge. This reached from the neck to two thirds of the length of the wing. The hood was

intirely

intirely of the ſame colour, except a large black ſpot in the middle near the neck. This hood had the ſame motions as the head, which was round, and ſerved it for an helmet: extending the breadth of a line beyond it, quite round. On the fore-part of the head, two black antennæ, as ſlender as a fine hair, and three lines in length, were placed above the eyes, which were black, not very prominent, and reſembled the ſeed of the amaranth. Three ſmall legs, equally black, came out from each ſide of the body; quite covered at the place of inſertion, with very ſhort, ſmall, fine hair of a yellow orange colour. The hind part was compoſed of five rings, the two neareſt the body, were black and covered with a ſhort velvet ſhag; the two next were inveſted with golden hair, and the fifth, which was as broad as either of the other two taken together, and which alſo completed the termination of the body in a rounded point, was likewiſe beſet with a black velvet ſhag, but rather longer than that of the other rings. Theſe rings were not ſo firm as thoſe which formed the body of the preceding fly; they bent under the ſlighteſt preſſure of the finger. The firſt fly emitted ſcarce any rays of light but from the head; this diffuſed them from every part of its body, the head alone excepted. Thoſe which I had kept cloſe in freſh graſs, lived four days, and preſerved the ſplendor of their light with almoſt as much brilliancy as at firſt, even to the time of their death.

Before we quit the road of St. Catherine's, I ſhall mention a few circumſtances relative to the Brazilians, which I learned from the firſt preſident of the ſupreme council of Rio Janeiro, whom I have ſpoken of above.

I did not ſee any bread made of corn, except at the Governor's of St. Catherine's. In every other houſe the Caſſavi bread was ſubſtituted in its room. This is a kind of a baked paſte made of the flower of the root of the Manioc, which is looked upon as a ſtrong poiſon, when it is eaten raw. I have however ſeen children, who were employed in taking off the rind of it to make the Caſſavi, eat it raw without finding any ill conſequence.

Some

Some of the inhabitants ufed to roaft it on the coals, then ftrip off the outfide and eat it.

There was but one fort of Manioc, which I met with on the ifland of St. Catherine's, and about the houfes on the continent. Laët, who is quoted in the collection of voyages publifhed by the Abbé Prevot, affirms that there are feveral forts of it, one in particular at Brazil, which is there called *Aypi*, and may be eaten raw without any danger. *Some nations*, adds he, *of the race of the Tapouyas eat likewife the common Manioc raw, which has the effect of poifon on all other people : but thefe are not hurt by it, being accuftomed to eat it from their infancy.* Thofe however, whom we faw eating it raw, were not of the race of the Tapouyas. They were white children born of Portuguefe parents. The leaves of this Manioc come very near in fhape to thofe of the Piony.

The roots are dried before the fire on hurdles, and then being fcraped with fharp ftones, are made into a kind of meal, the fmell of which refembles that of ftarch. This meal is put into large pots, where it is ftirred till it grows thick, as is done in France with the black wheat. When it is cold and become of the confiftence of a ftiff jelly, its tafte differs little from that of white bread. What they provide for their cruizes and land expeditions is thoroughly baked, and is therefore firmer and harder for the convenience of carriage. It is fometimes boiled in broth, and makes a very nourifhing difh, much like our rice-foup. Thefe roots pounded or grated frefh, and before they have been expofed to the fire, yield a juice as white as milk, which, if it is but fet in the fun, curdles like cheefe, and becomes very good food with the leaft affiftance of the fire. The method of fcraping the Manioc roots with fharp ftones, is an invention of the Brazilians, who are not at all acquainted with the mechanic arts of Europe. The Portuguefe, who were born or merely fettled on the ifland of St. Catherine's, and on the coaft of the main-land, which encompaffes it, ufe for this purpofe a large wooden wheel, the points of which have on the outer furface a groove. This groove is covered with an iron grater, upon which the

the roots are placed, one perſon preſſing them a little down, while another turns the wheel, which has the ſame effect as a tobacco grater. By this contrivance much time and labour is ſaved. But they did not preſerve the white juice, which dropped from theſe roots as faſt as they grated them, and falling into a ſmall hole, ran off upon the ground. After this operation the roots are dried in order to be reduced to meal, and to make the Caſſavi. But this is not their only uſe. The Brazilians make a drink of them. The proceſs in making of which is very diſguſting, as well as the liquor itſelf, to thoſe who know how it is made. The females are employed in this buſineſs, eſpecially the old women. Laët gives a particular deſcription of the whole.

The laws of every country form the manners of its inhabitants; hence it is, the manners of different nations have ſo little reſemblance to each other. The climate likewiſe contributes very much to produce this effect, it being evident that a law highly ſalutary in Norway may be equally pernicious in Guinea. The improvements introduced among thoſe, which we call *civilized* nations, have alſo given riſe to many laws unknown among, what we are pleaſed to call, *barbarous* nations.

Among the Brazilians, the girls before marriage not only give themſelves up freely, and without any ſenſe of ſhame, to unmarried men, but even their parents offer them to the firſt comer, and careſs their lovers exceedingly; inſomuch that perhaps there is not one girl who is a virgin at the time of her marriage. On the other hand, when they have once given their promiſe, which is the only ceremony on that occaſion, they are no longer ſolicited, neither do they liſten to any other addreſſes.

The only education they give their children is to make them expert in hunting, fiſhing, and war. They live peaceably notwithſtanding among themſelves, and very ſeldom have any private quarrels. If by chance any of them fall out and fight, they let them go on till they are ſatisfied; but as the law of retaliation is ſtrictly obſerved among them, the man, who has wounded another, is wounded in the ſame manner by his relations, or if he has killed him, is killed in his turn by them. All this is done by conſent of the relations on both ſides, and without appeal. This law is probably the ſource of that implacable hatred they bear to their avowed enemies. If this rule was introduced among us, we ſhould ſcarcely ſee ſo many quarrels ending in the effuſion of human blood; our only weapons then would be our tongues and our pens.

It is an injuſtice to the Brazilians to conſider them as the moſt cruel of all men againſt every other nation: they are only ſo to their avowed enemies: and excepting ſome few of certain nations whoſe ferocity approaches to that of wild beaſts, perhaps from the continued inſults of their neighbours, the Brazilians are very humane, particularly to ſtrangers, whom they receive with great kindneſs, and in the manner related by Lery; the particulars of which are as follow.

If one has occaſion to go oftener than once to the ſame habitation or village, one muſt take up one's lodging with the *Mouſſacat*, or head of the Family; becauſe the perſon to whom one goes firſt would be extremely offended at one's leaving him him to go to another. One muſt always lodge with the ſame perſon.

As ſoon as the traveller appears at the door, the Mouſſacat, or whoever is maſter of the houſe, preſſes him to ſit down on a hammock or bed of cotton, flung in the air, on which they leave him for ſome time without ſaying a word. They take this opportunity to aſſemble the women, who come and ſit on the ground round the bed, with their hands over their eyes. They melt with joy; they weep, and in the midſt of their tears addreſs a thouſand flattering compliments to their gueſt: "How good! How valiant you are! What obligations we have to you! What trouble you have undergone in coming hither! How beautiful you are! How happy you have made us by coming hither;" and

others

others to the ſame effect. If the ſtranger wiſhes to impreſs on them a good opinion of himſelf, he muſt appear to be affected in his turn. Lery affirms, that he has ſeen Frenchmen really affected and crying, as he expreſſes it, *like calves*. But he adviſes thoſe, whoſe hearts do not happen to be ſuſceptible of ſuch tender emotions (that is to ſay, to the ſhame of us, Europeans, who pique ourſelves, but with ſo little reaſon, on having more humanity than the Brazilians) to ſigh, or at leaſt pretend to ſigh. Is not this reproaching us in few words, with having only the outſide of politeneſs and hoſpitality, while the Brazilians are really polite and hoſpitable?

After the firſt ſalutation, the Mouſſacat, who has retired into a corner on pretence of making an arrow, or ſome other buſineſs, as if he was not taking any notice of what paſſed, comes near the bed, aſks the gueſt how he does, receives his anſwer, and enquires the reaſon of his coming. If the gueſt underſtands the language, he muſt anſwer all theſe queſtions. After this, if the *Mair* (for that is the name they give to Europeans) came on foot, they bring him water, and the women waſh his feet and legs. They then inquire if he is hungry or thirſty. If he anſwers, that he is both, they immediately ſet upon the table as much fiſh, fowl, veniſon, and other proviſions as they have in the houſe, with the liquors of the country.

Should the gueſt chooſe to paſs the night there, the Mouſſacat not only cauſes a fine white *inis* (or hammock) to be prepared for him; but, notwithſtanding the continual heat of the climate at Brazil, makes a pretence of the dampneſs of the night to kindle three or four ſmall fires round the hammock, which are kept up all the time the *Mair* is aſleep with a ſort of ſmall fan, called *Tatapecoun*, which reſembles very much our fire-ſcreens.

In the evening, ſays Lery, who had been in this ſituation himſelf, they remove their children out of the way; ſo careful are they to get rid of every thing that may interrupt the ſtranger's repoſe.

As ſoon as he wakes in the morning, the Mouſſacat comes to inquire if he has ſlept well, and how he finds his health; and though the ſtranger by his anſwer appear ever ſo well ſatisfied he cries, "Go to ſleep again, my child, you have occaſion for it: I am ſure you were very much fatigued yeſterday." It is cuſtomary with Europeans to make them ſome preſents upon theſe occaſions, and no man ought to ſtir abroad without carrying ſomething for that purpoſe about him. For this reaſon travellers provide themſelves with ſome trifling commodities, ſuch as knives, ſciſſars, tweezers (which laſt are in great requeſt, both men and women being accuſtomed to pull out the hair from all parts of their bodies, except their eyebrows) combs, little looking-glaſſes, bracelets, glaſs beads and buttons, and even fiſh-hooks.

Some doubts perhaps may be entertained in reſpect to the conduct of the Brazilians as repreſented in this account; but theſe will ceaſe, when it is known, that theſe men, whom we look upon as barbarians on account of their cruelty towards their enemies, never eat any other men but ſuch as are their declared enemies: that they bear a ſtrong affection to their friends, and allies; and that they would ſuffer themſelves to be cut in pieces, to defend thoſe with whom they live in amity from the ſmalleſt inconvenience.

It is not to ſtrangers alone that the Brazilians ſhew themſelves tender and affectionate. When they happen to be afflicted with diſorders, they treat each other with the moſt humane attention and regard. If any one is wounded, his neighbour immediately offers himſelf to ſuck the wound, and performs every other office of kindneſs with the ſame zeal.

Yet religion has no ſhare in regulating the conduct of the Brazilians. They have no notion of a Deity: they do not adore any thing, and their language has not even any word to expreſs the name or idea of a God. In their fabulous hiſtories, there is nothing to be found that has any relation to their origin or to the

the creation of the world. They have only a kind of ſtory, that ſeems to refer to the general idea of a deluge, which deſtroyed the whole race of mankind except one brother and ſiſter, who peopled the earth again. They unite ſome idea of power to thunder, which they call *Tupan*, for they are afraid of it, and believe, that they derive from it the knowledge of agriculture. They have not the moſt diſtant conception that this life is to be followed by another, and they have no terms to expreſs heaven or hell. It ſeems, however, that they have a notion of ſomething remaining of them after their death; for they have a tradition, that ſeveral of their people have been changed into genii, and dæmons, and that they rejoice and dance continually in charming fields planted with all ſorts of trees.

The Indians of Brazil are paſſionately fond of dogs of the European breed; and they bring them up for ſporting. Thoſe of the country, though they reſemble ours, are never broke of their ſavage and carnivorous appetite. A Portugueſe made us a preſent of two, one grown up, the other ſo young that he could ſcarcely walk. We were obliged to get rid of them both in time, for it was found that no correction could keep them from the ſheep and fowls. But the Governor had given M. de Bougainville a brace of pointers out of the ſame bitch, about four months old, and of the beſt breed in Portugal. When we were landed at the Malouines, and went out a ſhooting, they pointed naturally without any teaching. M. de Bougainville carried them to France, and made a preſent of them to a nobleman at court.

The 15th we ſailed out of the road, and continued our route the 16th and 17th, without meeting any thing remarkable.

The 18th, in the morning, we ſaw a great number of birds, which our ſeamen called *Dadins*, and ſome *Quebranta-hueſſos* or Oſpreys. One of the latter flying too near our veſſel was ſhot, and taken up out of the ſea.

It is an opinion which prevails in the South Sea, that the Quebranta-hueſſos never appear but a day or two before a ſtorm or very bad weather. But we ſaw many of them in fine weather, when it was what the ſailors call *a long ſea*, and no ſtorm followed. The ſame thing is ſaid of the Halcyons or King-fiſhers, or as they are otherwiſe called *Puants*. Whether it is that they really have a bad ſmell, or whether it is, that mariners have an averſion to ſeeing them, as they look upon them to be birds of ill omen, it is true we never ſaw any King-fiſhers but we had hard weather afterwards.

The Quebranta-hueſſos are frequently ſeen to ſtoop and hover on the ſurface of the water, ſkimming the waves, and following the direction of them, without appearing to move their wings, which they conſtantly keep open and ſpread out. When they do not float on the waves, they fly round the ſhip at a ſmall diſtance.

The body of this bird is about the ſize of a large capon, but its long cloſe feathers give it the appearance of being as big as a Turkey. Its neck is ſhort, and ſomewhat bent: its head large, and its beak very ſingular, ſuch as I have ſketched it out in the plate.

The beak is divided as it were into four or five pieces. The tail of the bird is ſhort, its back high, legs low, feet black and webbed, with three claws before, and a fourth very ſhort behind, each of which are furniſhed with black talons, which are blunt, and not very long.

There are ſeveral kinds of Quebranta-hueſſos. Some of them have a whitiſh plumage, ſpotted with a dark brown, or red; in others, the breaſt, the under part of the wings, the lower part of the neck, and the whole head, are all extremely white; but the back, the outſide of the wings, and the upper part of the neck, are of a dark red, ſpeckled with a few ſpots of a blueiſh grey colour. The one we ſhot was of this kind. They may perhaps be all of one ſpecies, and the differences may ſerve only to diſtinguiſh the male from the female. Their wings are very long. Seven feet two inches and a half was the length of our bird's wings, meaſuring from the extreme points

I of

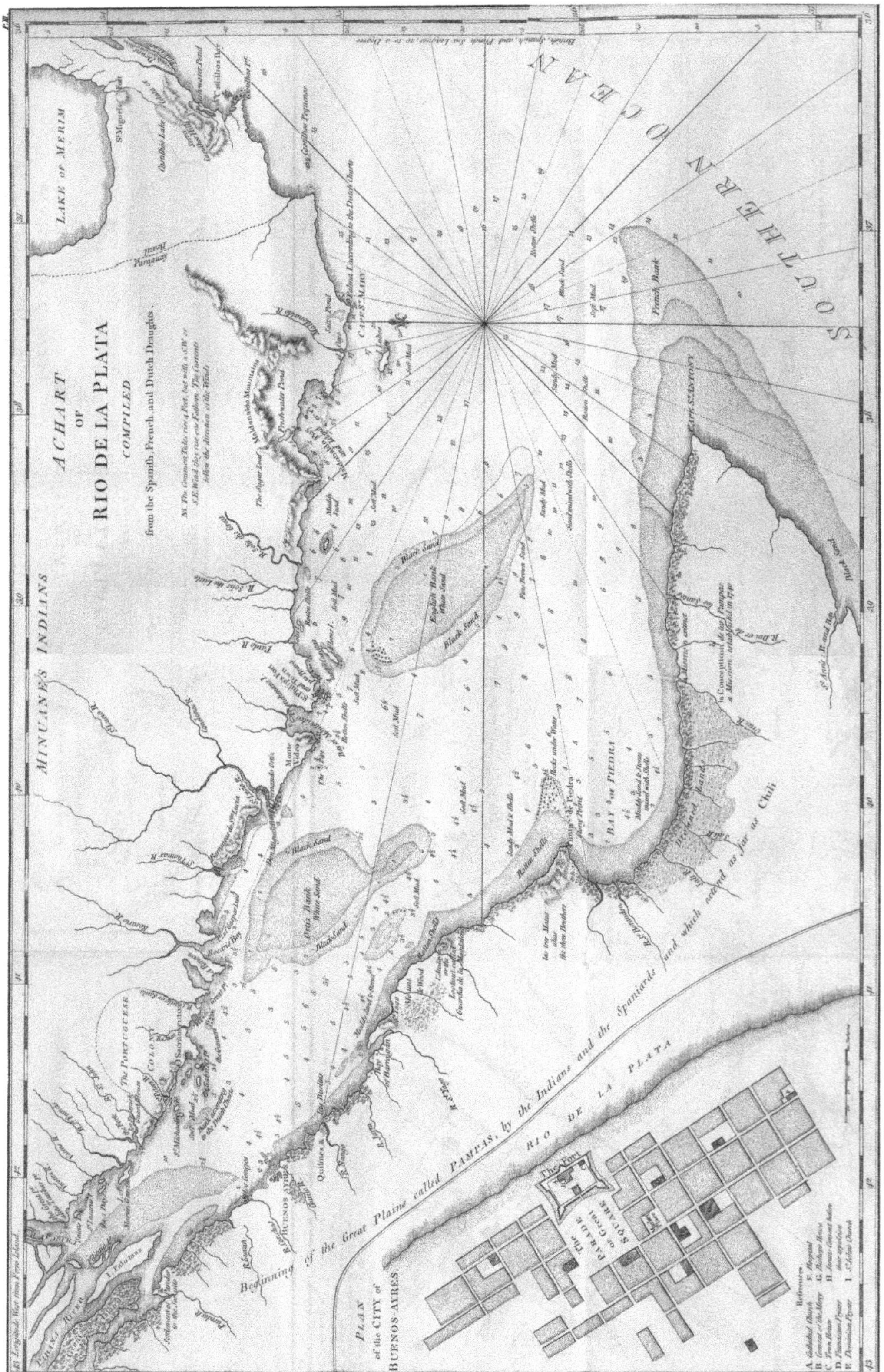

The material originally positioned here is too large for reproduction in this reissue. A PDF can be downloaded from the web address given on page iv of this book, by clicking on 'Resources Available'.

of the feathers of one wing to thoſe of the other. Theſe birds are found at the diſtance of more than 300 leagues from any land, neither is it known from whence they come, or in what places they build their neſts.

On the 21ſt in the morning, the colour of the ſea appearing different, we reſolved to ſound, and found bottom at the depth of fifteen fathoms. The end of the bank, which is at St. Mary's Point, may poſſibly make this bottom.

According to our obſervations, the currents from the line tend to the South when the ſun is in the Southern point.

On the 22d, about two o'clock, we ſaw land pretty clearly. We ſteered immediately towards it, in order to diſcover it well; the tides carried us S. S. E. from 14 to 15 minutes. This circumſtance muſt be attended to, in making the entrance of Rio de la Plata. The courſe ſhould be to the North.

As we came near the land, we judged it to be the moſt Eaſtern point of St. Mary's cape. Perceiving then ſome lands more to the South, we turned the head of the ſhip South Weſt ¼ W.: and at ſix o'clock diſcovered the iſland of *Lobos:* ſo called from its being inhabited only by ſea wolves, which are here met with in great abundance. As we came nearer to it we ſteered South ¼ South Weſt, in order to keep a league and a half out to ſea, that we might avoid a ridge of rocks Eaſtward of this iſland. This ridge extends near a league out at ſea. As it was dark, we did not perceive the mouth of the canal, which forms the iſland and port of the Maldonnados; ſo that we advanced nearly two leagues too far up the *Rio de la Plata*, or river Plata, in which we caſt anchor at eight o'clock in the evening.

On the morning of the 23d, our yawl went on ſhore with Meſſrs. de Bougainville, de Nerville, de Belcourt, l'Huillier, and Alexander Guyot, our ſecond Captain, to acquaint the Commander of the fort of the iſland of Maldonnado of our anchorage. They aſked leave to ſupply themſelves with freſh water and proviſions, which the Commandant gave them in the politeſt

politeſt manner. He was even ſo kind as to tell us, that we had not anchored ſo well as we might have done at the mouth of the canal leading to the port: that the anchorage at the port was by no means ſafe, and that we ſhould do very well to quit it; which we accordingly reſolved to do, as ſoon as the weather ſhould permit, and the Commandant ſhould ſend us a coaſting pilot.

The whole coaſt diſcovers nothing to the eye but low banks of ſand, ſome few remote eminences only appearing, which are called the mountains of Maldonnado, and are at the diſtance of ſome leagues from the coaſt. There are no trees to be ſeen, but a great quantity of cattle, very large oxen, and horſes. All the trade indeed of the country of Plata conſiſts in ſilver, and the hides of oxen.

On coming from the Eaſt, to enter into the Rio de la Plata, the iſland of Lobos is ſeen W. S. W. of the compaſs.

The weather had been calm, very fine, and very hot ever ſince the morning. Many of the crew employed themſelves in angling; and no ſooner was the line thrown into the ſea, than it was brought up again with a fiſh. Sometimes there were as many fiſh caught as there were hooks to the line. The fiſh were only of four or five different kinds. Some of them were what the Spaniards call *Viagrios*, and our ſailors *Machoirans*. The others were Carangues or *Guarencas*, dog and cat-fiſh of different ſizes, and ſome ſharks. We caught one cat-fiſh, one dog-fiſh, and two young ſharks.

The Machoiran, or beard-fiſh, has its belly flat, and ſome beards, as the little barbel; the head large, the ſkin covered with ſmall brown and almoſt imperceptible ſcales, nearly reſembling thoſe of the tench; at the root of the fins neareſt the head, there is a ſmall ſerrated bony proceſs, the teeth of which are inclined towards the body. This proceſs is as long as the fin, and has the ſame motions. When the fiſh wants to defend himſelf againſt other fiſh, or againſt the fiſherman, he puſhes out theſe proceſſes, and thruſts thrufts them into the bodies of other fifh, into the fifherman's hand, or, into any thing that is near him, even into wood if he can, to which he fometimes faftens himfelf by the ftroke. His puncture is venomous, fo that fifhermen take care to guard againft it. I know not whether there are any larger than thofe we caught. The biggeft of thefe was one foot and a half long, and four inches broad. It is a very well tafted fifh.

The Guarenca is an excellent flat fifh. We caught fuch a quantity of a kind of barbel fifh, that after the whole crew had been fupplied with it for this day and the next, the remainder of it was falted and dried, in the fame manner as the dried cod is prepared at Newfoundland. The largeft of thefe barbels, was of the fame fize in every way as the beard-fifh, defcribed above. The cat-fifh, and the dog-fifh, are kinds of fharks or fea dogs; they refemble them fo much in their figure that they are not eafily diftinguifhed: they were about two feet and a half long.

About three o'clock in the morning of the 24th, there arofe a violent South wind. At five o'clock the wind being fomewhat abated, we put ourfelves in readinefs to fet fail for Montevideo. The wind rifing again at feven o'clock, we caft our beft bower, by ten fathoms, on a muddy bottom of fine fand, and remained part of the day in this ftate.

M. de Bougainville, trufting to the mildnefs of the weather, had been gone ever fince the morning, to the Fort Maldonnado, with the fame perfons who had accompanied him the preceding evening; with a defign to view the country, as well as to procure frefh provifions. Fortunately for them they returned, between fix and feven in the evening. We had juft hoifted our anchor on the cat-head, on the profpect of a calm: no fooner was this done than a moft violent ftorm arofe, blowing from the South Weft part of the horizon. It is not poffible to behold a more beautiful fight, than that we enjoyed, from the continual numberlefs flafhes of lightning, which fhot from betwten the clouds as they rofe upon the horizon, which was all on fire; fo that fireworks the moſt elegantly contrived, the beſt kept up, and the moſt diverſified, could not poſſibly exhibit any thing to be compared to what we now ſaw in the Heavens for the ſpace of an hour. We did not then ſuſpect, that this proſpect would ſoon give way to one of a much leſs pleaſing kind; from which, our captain, who was better acquainted with the danger and the conſequences, was all this time endeavouring to ſhelter us, by taking all the neceſſary precautions againſt a ſtorm.

We thought the ſtorm would paſs off along ſide of us, as it appeared to be going that way; but in an inſtant a moſt violent wind aroſe, and the thunder and lightning came upon us at the ſame time. We paſſed the whole night ſtriving againſt the impetuoſity of the wind, and the roarings of an extremely boiſterous and angry ſea, which broke in upon us.

This wind is called in the country *Pamperos*, becauſe it comes from the plains of Pampas, beyond Buenos Ayres. Theſe plains extend as far as the Cordeleirias mountains, which divide them from Chili. They are at leaſt three hundred leagues in length, without the intervention of any wood, or eminence, to check the violence of this wind; which ſwells the river Plata, raiſing its waves as high as mountains, ſo as often to deſtroy the veſſels in that river, breaking them upon the coaſt oppoſite the wind. The anchorage where we were was extremely unſafe, from the vicinity of the iſland of Maldonnado, and the circumjacent coaſts, all bordered with rocks and ſands. Thirty years ago, an Engliſh veſſel laden with piaſters, or pieces of eight, was loſt in this place. The inhabitants of the iſland, in the neighbourhood of the ſpot where the wreck happened, are ſtill endeavouring to this day, to recover part of the cargo. On the evening before our arrival, they had taken up with the drag, two thouſand four hundred of theſe piaſters.

The wind Pamperos is much more frequent in winter than in ſummer, and always blows ſtrong, which makes the Rio de la Plata a dangerous road. This river is fit only for the trade

of

of piaſters and oxen, the largeſt of which are ſold here for five pieces of eight, or five and twenty livres of French coin. In general, they are ſold for three piaſters, or fifteen livres. It is very difficult to get wood here; not only becauſe it is very ſcarce, but alſo becauſe the little there is of it, is found a long ſide the rivers, the only places of ſhelter for tigers, leopards, and other wild beaſts, which are here found in great number, much more fierce and larger than thoſe of Africa and the Eaſt Indies. Figs and peaches are to be found all the way from Maldonnado and Montevideo to Buenos Ayres.

The Pamperos blew with equal violence the whole night; and the ſea was terrible. Notwithſtanding the continual rolling and pitching of the ſhip, I was fallen into a pretty ſound ſleep, when I was ſuddenly awakened by a violent ſhake the veſſel received, which made it crack in all its parts, as if it was breaking againſt the rocks. It was then near five o'clock in the morning. I jumped out of bed, opened my window, and aſked the helms-man if we had ſtruck againſt any rock. He anſwered, that we had not yet ſtruck, but that the ſhip dragged her anchors, and that we were in a fair way for it. He told me that the cable of our beſt bower had failed; and that the other anchor was looſened from the ground. This was the reaſon of the violent ſhake we had felt; our only hope was now in the ſheet anchor, which we had juſt dropped.

I dreſſed myſelf, and went upon deck, and I perceived indeed we had dragged ſo much, that the coaſts towards which the wind and the waves were driving us, did not ſeem to be more than half a league diſtant. But our ſheet anchor luckily keeping her hold, we continued in the ſame manner, labouring very hard for our preſervation, till about ſix in the evening, when the wind and ſtorm began to ceaſe.

As they were working the ſhip yeſterday evening, a pully broke. A ſailor received a ſtroke on the forehead from one of the ſplinters, and preſently after fainted away. The ſurgeons gave it as their opinion, that the ſtroke had only grazed the part, from the ſmall quantity of blood loſt, and the little apparent contuſion. The man was pretty quiet to-day, complaining of nothing more than a great head-ach, which was looked upon as the effect of the ſhock. He was blooded twice and very carefully attended, having even had a bed made for him in the cabbin.

During the tempeſt, the ſea was agitated quite to its bottom: two or three hours after the ſtorm began, the ſea formed ſuch deep cavities, that one would have thought we were going to touch the bottom; and the billows followed ſo cloſe that they did not allow us time to breathe. I ſaw more than once the end of the main maſt yard, which indeed was brought very low, dip three feet or thereabouts into the wave, part of which often broke upon the deck. Our ſituation was more dangerous from our being ſo near the coaſt. Our ſea officers, who were all able men, having had the command of ſhips, and privateers, were ſo ſenſible of the danger that threatened us, that moſt of them were conſidering how they ſhould ſave themſelves from the ſhipwreck. The danger appeared even ſo urgent to them, that the guns were already diſpoſed ſo as to ſerve inſtead of anchors, in caſe the cables ſhould happen to break.

The night of the 26th was fine, and the wind having ſhifted to the North, blowing a gentle gale, about four o'clock we began to prepare; and we ſet ſail about ſeven.

On the 27th at ſun riſe, the land neareſt to us bore N. N. E. about four leagues diſtant: and the land moſt to the South Weſt, bore N. W five degrees North of us. We ſounded the whole night from hour to hour, and even more frequently, and found ſometimes at twelve, ſometimes at thirteen fathoms, a muddy bottom. Thoſe who ſail along this road, will do well to ſound as often as they can, eſpecially if it is the firſt time of their going to Montevideo or Buenos Ayres. The river Plata is extremely dangerous, from the number and extent of its ſand banks, which leave but a very narrow channel, for the paſſage of ſhips, and that very ſerpentine. The bank called the Bank of of the English advances near five leagues from the coast; and the islands we meet with, form flats, projecting considerably.

All this coast is flat, except the part where the Maldonnado mountains are, which are of a moderate height, and at a small distance.

On the 28th we tacked about towards the island of Flora, till we came within about a league and a half from it, and till we were two leagues clear of the coast. One must take care not to go too near either of these places, on account of the ridges of rocks, which extend pretty near a league out at sea. At nine o'clock we were about two leagues distant from Montevideo. Half an hour afterwards we dispatched M. Alexander Cuyot in our yawl, to give notice of our arrival to the Governor.

About half an hour after four, as we were going to sail into the bay, the captain of a Spanish vessel, named Saint Barbe, came on board of us with offers of service from the Governor, and in order to pilot us. About five o clock we cast anchor in this road a little beyond the Spanish vessel, in three fathoms depth of water, on a muddy bottom; after which we saluted the citadel with twelve shots, which were returned shot for shot.

The first days we lay at anchor were taken up in settling with the Governor of Montevideo, what we had to do during our stay. He seemed first to start several difficulties, not only with regard to allowing us the liberty of fishing along the coast, but also to the suffering our longboat and yawl to come on shore. He required that we should give him previous notice every time we wanted to land, that he might post guards at the place where we should put to shore, with orders to wait till our long-boat or yawl went back, to prevent us from carrying on any kind of trade whatever.

Not foreseeing any of these difficulties, we had sent our small boat a fishing at the foot of the mount, the very next day, after we had cast anchor. The Governor who received intelligence of this, sent two dragoons of the garrison, to take the men into custody, together with the boat and goods, in case they should have

have brought any on ſhore. Meſſ. de Bougainville, de Nerville, Guyot, and myſelf, came into the palace, immediately after this order had been given, which the Governor communicated to M. de Bougainville. The Governor, apprehending certainly that he could not expreſs himſelf properly in French, ſpoke in Spaniſh, and had a provincial, ſettled in this city for fifteen years paſt, for his interpreter. This provincial explained the Governor's intentions to us in ſuch a manner, as induced us to believe, that he was not inclined to do us all the ſervices he had offered, and we had reaſon to expect from him. This however, was far from being his way of thinking; of which we were thoroughly convinced by the ſequel of our conference.

The order given, which ſeemed to agree with the provincial's interpretation, was far from being agreeable to M. de Bougainville, who expreſſed his reſentment of it to the Governor, nearly in the following terms: It is very ſurprizing, Sir, and at the ſame time very mortifying for us, to meet with difficulties among our friends the Spaniards, which we have not experienced among the Portugueſe, with whom we have juſt been at war. I will ſet ſail immediately, and give notice of this to the King my maſter. The Governor replied, that he did not intend to diſoblige us, but on the contrary, to do us all the ſervice in his power; that he was not the maſter: that the laws and orders of his court were, not to allow any kind of trade to be carried on by ſhips which were not Spaniſh, or privileged for that purpoſe by the court of Spain; and even to put a ſtop to any trade carried on by the Spaniards themſelves for other nations; that a frigate belonging to the Eaſt India Company, which had anchored in the ſame port three years before, had made no ſcruple of ſubmitting to what he had juſt propoſed. M. de Bougainville anſwered, there is a wide difference, between a trading frigate, and a King's frigate of war. We have no trading commodities on board; and are come here only with intent to take in ſome refreſhments, and to wait for the Sphinx, which we have parted from, and which we have appointed to meet in the Rio de la Plata.

Plata. As ſoon as you give me your word, replied the Governor, that no goods ſhall be diſſembarked; you are welcome to land or ſend on ſhore as often as you pleaſe. But the cuſtom eſtabliſhed by the laws, being to ſend a ſoldier or guard wherever the boat puts into ſhore, I deſire you would not take it amiſs that I comply with it; it is for your quiet as well as mine; for I do not chuſe to lay myſelf open to any blame from my court. In every other reſpect, you may depend on my doing every thing in my power to oblige you: for I am prompted by my own inclination, as well as by the orders I have received, to treat the French with the ſame civility as the Spaniards. Matters being thus ſoftened on both ſides, the dragoons were ordered to the boat, and went accordingly.

The Governor afterwards deſired M. de Bougainville to permit him to take a copy of the orders he had received from the king of France for the command of the two frigates, becauſe he was obliged to ſend it to the court of Spain, together with a circumſtantial account of our anchorage. M. de Bougainville readily complied with his requeſt: the reſt of the converſation was carried on amicably, and we parted good friends.

The Governor had more reaſons than one for acting as he did: he told us ſome of them, ſo that it was not difficult to gueſs at the reſt. Don Joſeph Joachim de Viana (which was the Governor's name) being now, in 1763, about forty-eight years of age, knight of the order of Calatrava, brigadier of the troops of his Catholic Majeſty, was intruſted by the king of Spain with the command of the troops ſent into Paraguay againſt the Indians, who had revolted, as it is ſaid, at the inſtigation of the Jeſuits. The Jeſuits then held the ſway in that country, and had refuſed to ſubmit themſelves to the meaſures taken by the courts of Spain and Portugal, to fix the limits of their reſpective poſſeſſions. Don de Viana, conducted this buſineſs, and all his operations were crowned with ſucceſs, notwithſtanding the obſtacles of all kinds which the Jeſuits threw in his way. This certainly was not the method to

8 obtain

obtain their good will, and indeed he knew that they were not his friends.

At Montevideo they have a convent, in which there are but two prieſts, and one layman, who as well as their emiſſaries, always keep their eyes open upon what paſſes, and are continual ſpies upon the conduct of the Governor of this city. The Governor of Buenos Ayres, whoſe power extends over the whole Paraguay, favours the Jeſuits in every particular, and ſcruples not to be their ſlave, and the mean inſtrument of their revenge. The Jeſuits knowing the miſunderſtanding there is between theſe two Governors, which perhaps they themſelves have excited, would not fail, if Don de Viana was in any ways capable of conducting himſelf in a cenſurable manner, to take advantage of any little circumſtance of that kind, and acquaint the Governor of Buenos Ayres with it: Don de Viana is thoroughly convinced of this. Being a man of great merit in every reſpect; a man of ſenſe, abounding in military knowledge, and diſtinguiſhed by probity; having none of that pride which the Spaniards are ſometimes reproached with, he has acquired the eſteem and reſpect of all who know him. All perſons are unanimous in his praiſe, from which even the Jeſuits themſelves cannot diſſent, at leaſt publickly.

There are more than ſixty Jeſuits in the convent of Buenos Ayres, which is ſaid to be very beautiful. The convent at Montevideo is but very ſmall, without any remarkable appearance, diſtinguiſhed from the houſes of the reſt of the inhabitants only by a bell, fixed under an arch, about three feet high, raiſed on one of the extremities of the top of the roof. I never ſaw the inſide, though the Jeſuits had ſent me invitations two or three times to come and ſee them. The provincial, already ſpoken of, firſt made the propoſal to me at the Governor's, and I came into it, promiſing to go the next day, or a few days after. A Spaniſh officer who was preſent told M. Bougainville of it, repreſenting to him that it was not proper a Frenchman ſhould pay a viſit to the Jeſuits, after what had lately happened at Buenos Ayres,

Ayres. He related the fact to Mr. de Bougainville, and taking me aside afterwards; you are a true Frenchman, said he, and you have just been promising to go and see the Jesuits. You must either break your word, or give up your pretensions to that title. About six weeks ago, one of them preaching at Buenos Ayres, when I was present, ran out in invectives against the Kings of France and Portugal, the Republic of Genoa, and against the other powers who have expelled the society: you cannot surely think of going to see them after this intelligence; besides, the visit would by no means be agreeable to the Governor. The last expression made me suspect that there was some partiality in this officer's discourse: so that I resolved to suspend my opinion. On coming away from the Governor', M. de Bougainville repeated this caution to me as coming from the same officer; I promised not to go near the society, and kept my word. The officer told me in particular, the invectives the Jesuits had used against the King of France, which were so remarkably indecent, that I have thought it best to pass them over in silence.

Two days after I found an opportunity of coming at the truth. I inquired about it of two Spanish officers, who spoke the French language well, and who were going to embark on board the Saint Barbe Frigate, in order to return to Spain. One of them was a colonel, the other a captain. The captain's name was Simoneti. They each of them confirmed the fact, and added, that as the Governor of Buenos Ayres made it a point to protect the Jesuits, he took no notice of this rash and impudent discourse; but that some persons of quality and distinction, whose probity was well known, had ordered a particular account of it to be made out, and sent to the court of Spain; and that they were the officers commissioned to carry a copy of it to the said court *.

Two

* These officers left Montevideo the same day that we did. The frigate upon which they embarked was commanded by Don Pedro de Flores, laden with 15 or 1,800,000 piasters, bulls hides, and other merchandize. She had set sail from Cadiz in 1755, bound to the coast of Guinea, fitted out on the English account, and destined

Two days after this converſation, I went to viſit the chaplain of a Spaniſh frigate, which had been lying at anchor in the port of Buenos Ayres for five months; I knew he was much inclined to the Jeſuits. It was even ſaid pretty openly that he had been ſent by them to Montevideo, in order to buy up any trifles he might find on board our frigate. He indeed bought up every thing we would ſell him.

After the firſt compliments had paſſed between us, he asked me why I had not been to ſee the Jeſuits, according to their invitation, and my promiſe. I told him it was true I had promiſed, but that I had been informed that one of thoſe fathers, had lately ſpoken very ill of the king of France, my maſter, in a ſermon preached at Buenos Ayres; and, if this was a fact, it was not proper that I, who was a true Frenchman, ſhould pay a viſit to the brethren of ſo raſh a preacher. You certainly heard the ſermon, ſaid I.——I did: and it is certain that the Father did not expreſs himſelf in the moſt cautious manner.—What ſaid he then particularly of the King of France?——That he was a tyrant, a perſecutor of the church, and many other things. But we ſhould ſurely forgive them, for this is nothing more than the effect of their reſentment having been expelled the kingdom of France.

We had ſcarce finiſhed this converſation, when two of the three Jeſuits of Montevideo came into the room where we were, the Abbé, M. Mauclair our ſurgeon, and myſelf. After having bowed to us, one of the Jeſuits, addreſſing himſelf to me, ſaid he was very glad to ſee me; and that in conſequence of what he had heard from Joſeph (this was the name of the provincial before mentioned) he and his brethren had expected me for two days. He afterwards aſked me why I had not kept my word; I anſwered that

deſtined to carry over negroes to Buenos Ayres; but not meeting at Cape Verd with the Engliſh veſſel from which ſhe was to have received them, Don Pedro de Flores continued his courſe, and ſailed into the Rio de la Plata. Here he had remained ever ſince that time, in order to avoid running the riſk of being taken priſoner by the Engliſh during the laſt war, as he was laden for Spain. From the obſervations he had collected for making a chart of this river, joined to my own, the chart inſerted in this work has been traced.

that I had juſt been giving my reaſon to the Abbé, who might explain it to him. I am not ſurpriſed, ſaid he, I know the Benedictines do not think properly, and that they are not our friends. I told him he was miſtaken, and that if they did not think properly, they would be his friends. As he did not like my anſwer, he made no reply; but took his leave of the company, and went away with his companion.

Between four and five in the evening, we ſpied a ſail. We judged immediately by her courſe, that ſhe was making for Buenos Ayres. But as we expected from day to day the arrival of the Sphinx ſloop, which we had appointed to meet in the Rio de la Plata, many of us imagined this was ſhe. As ſhe advanced, and came more within our notice, we were ſo far confirmed in our opinion, as almoſt to perſuade ourſelves that we were aſſured of it. Notwithſtanding the uncertainty, M. de Bougainville diſpatched the long boat with the lieutenants, Donat and Le Roi, to pilot her. The ſignals were agreed upon, powder and other neceſſary articles were given to put them in execution, and they ſet off about ſeven o'clock. The night grew very dark, the winds contrary, and the ſea rather high, ſo that not having perceived their ſignals, we grew very anxious about them. The Sphinx had diſcovered us by the ſignals agreed upon, and in order not to loſe ſight of us, had done nothing but ply to windward, and make ſeveral tacks, which together with the darkneſs had prevented our long boat from boarding her: this ſhe did however at midnight. The Sphinx then anchored, and ſetting ſail the next morning, being the firſt of January, ſhe came up, and caſt her anchor near us about nine o'clock in the morning. The joy we had at ſeeing her, after a ſeparation of more than two months, may readily be conceived. M. de la Gyraudais had been previouſly told of the defect in the maps, with reſpect to the bearing of the coaſt of Brazil; but though we had ourſelves been upon our guard, we were very near running aground upon the bank which is not marked in the French charts. This bank lay in his courſe as it had done in ours;

nor are the Abrolhos made to extend ſo far upon the charts as they really do: all theſe circumſtances contributed to make us uneaſy, on account of her delay, eſpecially after the ſtay we had made at the iſland of St. Catherine.

As ſoon as the Sphinx had caſt her anchor, M. de la Gyraudais came to us in our long boat, and told us he had been obliged to put in at Togny on the coaſt of Brazil, becauſe notwithſtanding they had been appriſed of the errors in the charts, yet they had fallen upon the Abrolhos at a time when they thought themſelves at leaſt thirty leagues diſtant from them. They found themſelves ſtuck upon them in the middle of the night; it happened luckily that the weather was calm, and that the rock upon which they ſtruck was of rotten ſtone.

The Sphinx being faſt upon this rock, in order to avoid the dreadful conſequences of a wreck, they quickly hoiſted out the fiſhing boat they had, put the long boat and the yawl to ſea; and having carefully examined the ſhip, they recovered a little from their apprehenſions, when they found ſhe had received no damage.

The next trouble they had was to diſengage the Sphinx from the rock: as ſoon as it was day light, they found themſelves ſurrounded with rocks of the ſame kind; and at the diſtance of half a quarter of a league, a veſſel lying on her ſide, without maſts. M. de la Gyraudais imagining they were then upon the Abrolhos, and that land could not be far off, ſent the boat towards the ſhore for aſſiſtance. They met with ſeveral canoes of fiſhermen, negroes and Indians. They ſpoke to them in the Portugueſe language, and ſix of them agreed to go on board the Sphinx, where they were well feaſted. They promiſed to give them all the help in their power. Two of them were kept on board, and the other four diſpatched in the boat, to bring up their comrades from the coaſt. They came back the next day attended by a great number of canoes. With their aſſiſtance, the Sphinx was at laſt diſengaged from the rock, after having reſted upon it for three days. M. de la Gyraudais came

off

off with the loſs of the fiſhing boat only. The negroes piloted him as far as Togny, where the inhabitants treated him and his crew, for ſix days, with the greateſt humanity, and as well as if they had themſelves been of this country: although they are moſt of them negroes or Brazilians.

After this interview, we went to Montevideo to pay our compliments to the Governor upon the new year, not knowing that it was cuſtomary in this country to defer this ceremony to the ſixth day of the month, the Epiphany. The Governor was holding a council for the nomination of officers of juſtice. Being informed that, after the finiſhing of this buſineſs, he was to go with all the retinue to the pariſh church, which they call the cathedral, we went to the ſpot, and waited for him upon our legs a whole hour, under the ſhade of a houſe, converſing with ſome officers of the garriſon. At half an hour after twelve, he made his appearance in the midſt of the new officers of juſtice, who had each of them large white wands in their hands, which they made uſe of, as walking ſticks. He croſſed the ſquare, which is very large, in the middle of theſe officers, all ranged in one line, having their large black cloaks on, and their ſmall ſticks; as the Oviodore of the iſland of St. Catherine. We followed them into church. Maſs was performed by the prieſt, whom they called *le Signor Vicari*; when this was over, we paid our compliments to the Governor, who invited us to dinner. As we were already engaged to dine with him the next day by appointment, M. de Bougainville thought proper to decline the acceptance of this kind invitation, and went on board with M. de Nerville.

I ſtayed behind in expectation of dining with the vicar; M. Duclos our captain having told me the evening before that I ſhould do this gentleman a great pleaſure, and that they had talked about it. After having ſaluted the governor, I went up to the vicar in the veſtry, but did not ſpeak a word to him about dinner. We came out of church with the two Spaniſh officers, who were to embark on board the St. Barbe; we went along with the vicar a little way without receiving any invitation to dinner;

dinner; and I took care not to invite myſelf. When we had left him, the captain aſked me where I ſhould dine. I anſwered that I did not know; that I had expected to dine with the vicar; but as he had not mentioned any thing about it, I intended to ſeek my fortune ſomewhere elſe. He immediately ſaid, that I ſhould go with the colonel to the Governor's. I ſtarted many objections to this, not thinking myſelf ſufficiently known to go in this manner; the colonel inſiſted, and taking me by the hand, told me the Governor would be pleaſed with him for bringing me; and that he certainly would take it amiſs if he ſhould know that I had refuſed. I conſented therefore, and was received by the Governor and his lady, with all poſſible marks of politeneſs and favour. He ſpeaks French well enough to be underſtood; his lady underſtands it without being able, or rather without venturing to ſpeak it. Her huſband and the colonel were her interpreters. She was a native of Biſcay, tall, well made, of a brown complexion, but her features were rather too maſculine. She is a woman of great wit and vivacity, and about thirty-four or thirty-five years of age.

At eleven o'clock on Monday morning the ſecond of January, M. de Bougainville, Meſſrs. de Nerville, de Belcourt, l' Huillier, the two Du Clos brothers, our firſt and ſecond captain, Donat our firſt lieutenant, de St. Simon a Canadian, lieutenant of infantry, de la Gyraudais, captain of the Sphinx, and myſelf, all went to the Governor's, where we had as elegant a dinner as the country would afford; but the diſhes were dreſt according to the cuſtom of the place: that is to ſay, moſt of them with the fat of oxen clarified which they uſe inſtead of butter and oil; and ſeaſoned with ſuch a quantity of pimento and carthamum that the victuals were quite covered with them. Care had been taken however not to put theſe ſpices upon all the diſhes, and many of us eat of none but theſe laſt. The only wines offered us were Spaniſh, and wines from the country of Chili; the plates and diſhes were ſilver, and ſome of them china. The table was covered with a very ſhort cloth, and the napkins were rather leſs

I than

than handkerchiefs of a moderate ſize, naturally fringed, or, to ſpeak more properly, unravelled at both ends. The diſhes were ſerved up one after another. When drink was called for, it was neceſſary to ſay, whether one choſe wine or water, or a mixture of both; for the Spaniards generally drink nothing but water at their meals: at the end of which it is cuſtomary to bring a large glaſs of wine to every body in company, even without its being aſked. When wine and water was called for, they were brought one after another, and we were obliged to drink them ſeparately. The wine of Chili is of the colour of phyſic, compounded of rhubarb and ſenna, and very much of the ſame taſte. It takes this taſte perhaps from the ſoil, perhaps from the goat ſkins lined with pitch, in which it is conveyed. There is ſcarce any other wine drunk in Paraguay. One ſoon accuſtoms one's ſelf to this taſte; and after having drunk it for a few days, one finds it good. It is very warm upon the ſtomach. But, whether from taſte or fancy, the Spaniards preferred the wine we had brought with us from France. The deſert was entirely compoſed of ſweet-meats. The bread though made with excellent flour, was not good, becauſe it was not well leavened nor properly kneaded; neither do they know how to bake it.

In the evening M. de Belcourt, who had taken a lodging in the town, met with a ſtranger in company, perhaps in diſguiſe, who ſpoke a gaſcoon French. Prompted in all probability by the Jeſuits, who had already taken care to acquaint themſelves by the people belonging to our frigates, of M. de Belcourt's military reputation; this man propoſed to him to enter into the ſervice at Paraguay, in order to form the troops. He made him promiſes from the Jeſuits, of the higheſt emoluments to induce him to accept of the propoſal. M. de Belcourt pretended to liſten to him, but without entering into any engagements; and the very next day acquainted M. de Bougainville with this circumſtance. This gentleman anſwered, that ſome political advantage might poſſibly be made of this, and that if he choſe to ſacrifice himſelf for the good of his country, it might then be proper to give

give an ear to theſe propoſals. M. de Belcourt anſwered, that in caſe he ſhould think of engaging in this buſineſs, it would be neceſſary that M. de Bougainville ſhould give him a note, certifying that he went with his conſent, and for the preſumptive good of the ſtate.

The next day the ſtranger renewed his ſolicitations to M. de Belcourt with greater earneſtneſs, deſiring him to take his reſolution ſpeedily; that he need not trouble himſelf about his cloaths or any thing elſe; that care ſhould be taken to ſupply him with every thing he might want; and that, in order to prevent the Spaniſh government from knowing any thing of the matter, he ſhould be conducted, by ways unknown to the Spaniards, to the place of his deſtination. M. de Belcourt inquired which was the place, and what were the advantages propoſed; but the ſtranger not giving any ſatisfactory anſwer, and having talked to him in a ſlighting manner of the Jeſuits, on purpoſe to conceal his deſigns more effectually, M. de Belcourt declared at once, that he would not comply with his ſolicitations. But as he was under ſome apprehenſions how he ſhould get away, he kept himſelf on his guard. About the duſk of the ſame evening, he found himſelf ſo cloſely preſſed by three men, that he thought himſelf obliged to draw his ſword, and carry it out of the ſcabbard, to make his way, in caſe they ſhould have ſurrounded him; which, however, they did not attempt. I had all theſe circumſtances from his own mouth, and it is with his conſent I make them public.

Towards eight o'clock in the evening, M. Mauclair, firſt ſurgeon of our frigate, came and told me, that after having had a conſultation with M. Baſle the ſecond ſurgeon, and M. Frontgouſſe ſurgeon of the Sphinx, upon the preſent ſtate of the ſailor who had been wounded in turning the cap-ſtern during the late ſtorm, they had agreed that he was growing much worſe, and that he himſelf deſired to be confeſſed. I went down immediately, and finding him indeed very ill, received his confeſſion.

He

He loſt his ſenſes an hour after; about ten o'clock I adminiſtered the extreme unction, and at eleven he died.

On Tueſday morning the 3d inſtant, having previouſly apprized the vicar, we ſent away the corpſe in our yawl. It was depoſited with the guard of the port, till the vicar ſhould come to meet it. He came an hour afterwards, with his ſexton. On his arrival, I made him a compliment in Latin, to which he gave no other anſwer, but a very low bow. He had a Roman ſurplice on, and a gown; his ſexton, a lay-man, had a black petticoat on by way of caſſock, and a very dirty ſurplice. Meſſ. Duclos, Guyot, his brother Alexander, his two ſons, ſix ſailors, and myſelf attended the proceſſion. At each turning the vicar chanted a reſponſe, and a prayer, and ſang alſo the maſs for the dead. He did the ſailor all the honours he could have done to the captain himſelf, and had him buried in the church. The ſervice being over, he invited us to dinner, and could not be prevailed upon to accept of any fee.

After dinner, I took a walk towards the extremity of the creek which forms the port, where our people were getting water. I went all over the coaſt and the adjacent ſoil, in expectation of finding ſome curious plants or ſhells; but my ſearch was fruitleſs. I met with one ſingle plant only in a ſtate of perfection; the ſtem, which is eight or ten inches high, and the leaves were covered over with a ſhort white down, ſo cloſe and ſo thick, that it concealed the green part from the ſight. I am unacquainted with the name and properties of this plant.

At the diſtance of two ſhots or thereabouts from the creek or bay, there are two fountains. The people of the country waſh their linen in that which is neareſt the river. It is forbidden to waſh in the other; becauſe that is the one from whence they draw the water uſed for drink in the town, which is at the diſtance of half a league from it. This fountain is bordered with a little wall of ſtone, and is very badly kept up, though at the king's expence; ſo lazy are the inhabitants, and ſo careleſs even of what concerns them nearly.

As I was paſſing by this fountain, I ſaw three or four Mulatoes, who had brought there ſome ſtones upon a cart, drawn by four large oxen; and three others, who were filling a caſk with water, in order to carry it into the town. An Indian or Mulatoe woman, with a negro woman coming to the ſame place to draw water alſo, one of the Mulatoes, who looked very much like an Indian born of Spaniſh parents, took the negro woman by the hand, and they both of them danced together upwards of a quarter of an hour, the dance called Calenda. Travellers who ſpeak much of this dance in their accounts, do not exaggerate, when they deſcribe it as the moſt laſcivious of all dances, at leaſt judging of it by our manners.

It is thought, that this dance has been brought into America, by the negroes of the kingdom of Arda, upon the coaſt of Guinea. The Spaniards dance it as well as the natives, throughout all their eſtabliſhments in America, without making the leaſt ſcruple about it; although the dance is ſo very indecent as to aſtoniſh people who are not uſed to ſee it. It is ſo univerſally, and ſo much liked, that even children, as ſoon as they are able to ſtand, imitate in this particular perſons more advanced in life.

It is danced to inſtrumental as well as vocal muſic, by two or by ſeveral perſons together. They are all diſpoſed in two rows, one before the other, the men oppoſite to the women. Thoſe who grow tired, as well as the ſpectators, form a circle round the dancers, and the muſic. Some one of the dancers ſings a ſong, the chorus of which is repeated by the ſpectators, with clapping of hands. All the dancers keep their arms half raiſed up, jump, turn round, make contorſions with their backſides, advance within two feet or thereabouts of one another, then fall back in time, till the ſound of the muſic or tone of the voice brings them together again. Then they ſtrike their bellies one againſt another two or three times following, and retreat afterwards, whirling about, to begin the ſame motion over again, with jeſts, which are extremely laſcivious, indicated by the ſound of the inſtrument or voice. Sometimes they mix their arms, turning round two or three times, and continuing to ſtrike themſelves upon the belly, and to kiſs each other, without being in the leaſt out of time.

One may readily judge, how ſurpriſing ſuch a dance muſt appear to French manners, and how much our modeſty muſt be offended by it. Neverthleſs we are aſſured from the accounts of travellers, that it is ſo very agreeable even to the Spaniards of America, and is become ſo much an eſtabliſhed cuſtom among them, that it is even introduced among their acts of devotion: that they dance it in church, and in their proceſſions: that even the nuns themſelves, ſcarce ever fail to dance it on Chriſtmas-eve, upon a ſtage raiſed up in their choir oppoſite the grate, which is left open, that the people may partake of the ſight; but they do not admit men to dance with them.

On Wedneſday the fourth of January, while Meſſ. de Bougainville and de Nerville were gone to the Governor, to invite him to dine on board our frigate for the Sunday following, I went to ſee an officer whoſe name was Belia, who had been brought up in France, in our royal college of Pontlevoy near Blois. He had promiſed me ſome curious and medicinal plants of the country, and ſome pieces of natural hiſtory. With reſpect to the laſt article he had nothing worthy of attention; but he ſhewed me the plants, which I ſhall now deſcribe: his brother-in-law and himſelf acquainted me with their names, properties, and uſes.

One called *Mèona*, is very much like the wild thyme, but the leaf is round, and the green not ſo dark; the ſtem red, creeping, taking root at each joint, affording a white milky juice, like the ſpurge. The ſeed grows in a ſpiral, briſtly pod; this pod contains only a yellowiſh ſeed in form almoſt like a kidney. It throws off from its root ſeveral woody ſtalks, which ſpread themſelves circularly on the ground, as thoſe of the biſtort. This plant taken in infuſion, like tea, is ſaid to cure a ſtoppage of urine as by miracle.

Ebreno, or *Mio-mio*, is an almoſt repent plant, not riſing more than half a foot from the ground. The leaf is ſmaller than fennel, it has a very ſmall herbaceous flower, growing in cluſters, and pretty nearly umbellated: the root is reddiſh outwardly, and as well as the plant has the taſte of the parſnip. It is taken in infuſion againſt fluxions and colds. It ſeems to me to be a ſpecies of the *Meum*, or ſpignel.

The *Maté* has a round ſtrait branched ſtem, growing about a foot and a half high, and covered with a grey down a little inclining to red. The leaves are an inch and a quarter in length, only three or four lines in breadth, of a whitiſh green colour, and downy on the ſtem. The flowers ſhoot out one by one along the branches, and are compoſed of a ſingle yellow leaf, ſlit into four, and almoſt without ſmell. They are ſucceeded by a huſk or pod, of the thickneſs of a quill, an inch in length, which opens itſelf into four parts when dried, and lets fall ſome exceedingly ſmall ſeeds pointed at each end, of a grey brown colour. It is ſaid to be of admirable efficacy when applied to wounds, either recent or of long ſtanding. M. Simoneti told me, that, after having been ſix months under the care of the phyſicians and ſurgeons of the army, for a wound he had received in the ſide near the kidneys, and which had degenerated into an ulcer, he had cured himſelf in a ſhort time merely by the outward application of the leaves of this plant.

The *Cacken-laguen* or the *Canchalagua*, which is alſo called at Chili, *Cachinlagua*, is in every reſpect like the leſſer Centaury of Europe. It is the Centaury of Chili, but does not grow quite ſo high as ours. A cold infuſion is made of it, by throwing ſix or ſeven of the plants whole and dry into a glaſs of water for the ſpace of the whole night, or from morning to evening. This infuſion is then uſed as a gargle, and afterwards ſwallowed, by which method a ſore throat is ſoon cured. Some freſh water is then poured upon the reſiduum, which is ſuffered to ſtand as long as the firſt; after which the gargling and deglutition is repeated. This is done alſo a third time. M. de Bougainville, and M. du Clos

Clos our captain, had experienced the efficacy of it more than once. When the infuſion is taken warm in the manner of tea, it heats very much but purifies the blood. This plant is very famous in Chili, from whence it is brought. I believe it to be a better febrifuge than the Centaury of Europe. Might not the latter be uſed with equal advantage in ſore throats?

Mechoacan, is a name the Spaniards of Montevideo give to a plant bearing no reſemblance to that which is ſold in our ſhops under the ſame name. That of Montevideo, which is very common there, as well as in the neighbourhood of Buenos Ayres, is a ſmall creeping plant, the root of which runs under ground like the liquorice. It is whitiſh, and ſlender as a writing pen; ſome ſhort branches ſhoot out from this root, which creep upon the ground, are covered with a very few ſmall leaves, and theſe only at the extremity, almoſt reſembling thoſe of the leſſer Tithymalus, known in ſeveral provinces of France by the name of Réveil-matin. M. Belia told me, that the Engliſh who trade at the colony of St. Sacrament, always carry away ſeveral of theſe roots. It has a purgative quality like the Mechoacan of our ſhops. When it purges too violently, its effect is ſoon ſtopped, only by ſwallowing a large ſpoonful of brandy.

Another plant which they hold in great eſteem is the *Guaycuru*; it bears a leaf of a beautiful green colour, rather thick, and ſhooting forth in great abundance from the root, which is of a red brown colour, externally ſhining, and reddiſh within, as the ſtrawberry plant. From the middle of the root, the ſtem grows out to the height of half a foot, of the thickneſs of a common quill, ſolid, without leaves, of a greyiſh coloured green, ſpreading out at the upper part into a dozen ſmall branches, bearing at their extremity very ſmall herbaceous flowers, without ſmell, and forming altogether a kind of umbrella.

This plant, eſpecially the root, is one of the moſt powerful aſtringents in botany; and experience has proved, that it never fails in drying up and curing ulcers ſpeedily; and even, as the vicar told us, in curing the ſcrophula, and ſtopping a dyſentery.

He

He made us a preſent of a dozen of the plants which he had ſent for on purpoſe, at the diſtance of a few leagues, from a country place belonging to him.

The *Payco* is a plant, which throws out from its root ſeveral creeping branches: theſe are afterwards ſubdivided into many others. The leaves are but three lines in length, and two in breadth, ſerrated, thick, and fixed to the branches without any root-ſtalk. The flower is ſo ſmall that it is confounded with the ſeed, which ſucceeds it, and with which the branches are almoſt entirely covered. At firſt ſight, it might be taken for the *Rupture-wort*, or *Herniaria*, if the branches were ſhorter. The whole plant is of a pale green colour, ſometimes reddiſh, as well as the ſtem, when it approaches to maturity. It ſmells like a lemon juſt beginning to ſpoil. It is an excellent remedy for diſorders of the ſtomach, and indigeſtion. Its decoction is ſudorific, and its virtues are much extolled in the pleuriſy. The method of taking it, is, by chewing one of the green ſtems about the ſize of one's little finger, and ſwallowing afterwards the ſaliva together with the chewed plant. When taken in this manner it is a mild purgative. When there is none of the green plant to be had, it is taken in infuſion like tea.

M. Belia ſpoke highly of the anti-venereal virtues of the *Colaguala*, which ſome call *Calaguela*. It grows in barren and ſandy ſoils, to the height of ſeven or eight inches. Its ſtem conſiſts of ſeveral ſmall branches, which ſhoot up through the ſand or gravel. They are but two or three lines in thickneſs, full of joints placed at ſmall diſtances from each other, and covered with a pellicle which falls off of itſelf when it is dry. The leaves are very ſmall, few in number, and ariſe immediately from the ſtem.

The colaguala is looked upon as an admirable ſpecific for diſſipating impoſtumations in a ſhort time. Three or four doſes, that is to ſay, three or four pieces of it in ſimple decoction, or infuſed in wine, and taken in the courſe of the day, are ſufficient to effect this purpoſe. Being a very hot plant, it would become injurious

injurious if taken in too large a quantity. The root, which is the only part of the plant in uſe, is of a reddiſh brown colour outwardly, and reſembles much the *Guaycuru* root. When cut horizontally, it has a brown ſpot in the center, and a whitiſh circle in the middle of its ſubſtance. A Franciſcan named Father Rock, famous for his knowledge in phyſic, told me, that he preſcribed the *Colaguala* in the epilepſy, as well as in the venereal diſeaſe; that when it did not ſucceed perfectly in the cure of the epilepſy, he had aſſiſted it with the following preſcription, which had never failed of ſucceſs. He makes the patient drink, in the courſe of the day, a quart of water, in which a young virgin arrived at the age of puberty, or a ſound healthy woman, has well waſhed the parts of generation on getting out of bed; with particular directions that two glaſſes of this water ſhould be taken faſting, one half an hour after the other This remedy is continued for eight or nine days conſecutively, at the decline of the moon; and is repeated for ſeveral months, eſpecially in the ſpring. The method of uſing the *Calaguala*, in venereal diſorders, is by infuſion in wine, or in boiling water.

The ſame Franciſcan being with us at the Governor's country houſe, ſhewed me another plant which he called *Carqueja*; and which he told us was admirable, in infuſion like tea, for diſſolving coagulated blood in the body, for purifying it, and removing obſtructions. But it muſt be uſed very ſparingly, as it agitates the blood violently, eſpecially the root of it.

The *Carqueja* grows like a ſmall ſhrub, to the height of one foot, and its head is naturally rounded. It has no leaves diſtinct from the ſtem, which reſembles much that of the Geniſta or broom, with which I fancy it may be claſſed. This ſtem divides itſelf into many branches to form the head. Theſe branches are very flexible and thin.

The *Yguerilla*, the *Zarca*, and the *Charrua*, are plants greatly valued in this country; as well as the *Birabida*, or *Viravida*, which is reckoned refreſhing and cooling in the higheſt degree. A French ſurgeon preſcribed an infuſion of the Birabida with good ſucceſs

ſucceſs in a tertian. Frezier reckons it among the ever-greens. May it not probably be the ſame as I mentioned before under the name of Doradilla?

But the plant they make the moſt uſe of is the *Séfran.* It is properly a kind of thiſtle, known under the name of Carthamum. The deſcription of it is found in every botanical treatiſe. Its flower is called the baſtard ſaffron. It has the colour and form of the true ſaffron; but has not either its taſte or ſmell. At Montevideo and at Brazil they ſow the *Séfran* plentifully in their gardens; becauſe they uſe the flower of it to cover all their victuals, and even the ſoup. Parrots and Paroquetes are very fond of the ſeed, which is white, ſmooth, and made like that of the Corona Solis, or Sun-flower, but much ſhorter.

M. de Bougainville having told me, before he went to invite the Governor, that we ſhould ſet out early to go on board again, I went to the yawl at half an hour paſt four. There I found M. de la Gyraudais, and the ſurgeon of the Sphinx. After having converſed ſome time about the plants I had been collecting, finding that M. de Bougainville did not return, M. de la Gyraudais propoſed taking a walk about a mile off, behind the citadel, telling us, that the plant Maté had been ſhewn him; and that there was a great quantity of it near a fountain.

M. Frontgouſſe, who had alſo heard of its properties, came with us in order to gather ſome. We collected likewiſe ſome of the ſeed, which I gave, as I did all the ſeeds I collected in the courſe of my voyage, to M. de Juſſieu, to ſow them in the King's Garden at Paris. While we were ſupplying ourſelves with this plant, we heard a plaintive ſound iſſuing from between a large heap of ſtones and rocks, which cover and ſurround the fountain: we were not more than ſeven or eight toiſes diſtant from the ſound. We thought at firſt it proceeded from a cat confined among theſe ſtones, which might have eſcaped from a houſe about half a mile diſtant. As we came nearer the fountain, the cry ſeemed like that of a child. We were advancing towards

towards it, when M. Frontgousse desired us not to proceed, saying, it was not the cry of a child, but that of an alligator. He told us, he remembered to have heard them more than once in our islands, and that had we proceeded it would have been to our cost. We found indeed that there were alligators in this country; M. de St. Simon having already told us, he had seen one of them on the bank of a small river, running behind the mountain, separated from the town only by the bay in which the port is situated. Not daring therefore to push our curiosity any farther, we contented ourselves with gathering a few more plants, and went back towards the town, in order to go on board again. As we were walking along, we met with several Curlews, by thirty in a body. They came within pistol shot of us, but we had only sticks in our hands.

About seven o'clock we reached the yawl, where we met with Mess. de Bougainville, de Nerville, de St. Simon, and Martin, Lieutenant of the Sphinx. It was very fine weather when we left the Port; and we had already made three parts of our way, when a South East wind arose so briskly as to oblige us to ply our oars, in order to get on board, before it should become more violent. It blew however harder and harder. Each cloud as it rose on the horizon brought a fresh squall, more violent than the preceding. The waters being considerably swelled by these repeated attacks, formed waves which grew bigger and bigger, and retarded our progress. Although the sea and the wind were against us, we were now within gun shot of the Sphinx, which was the nearest vessel, and on board of which we thought of setting M. de la Gyraudais, with the other officers belonging to her. The fine clear sky had disappeared. The clouds made the night still more dark, so that we could but just discern the figure of a boat, bearing towards us. We then imagined that M. Duclos, suspecting our distress, had sent out the longboat to our assistance. We haled her, but received no answer. The sea however drove her towards us with so much swiftness, that we soon discovered her to be our small boat, floating at the mercy of the waves, with no perſon in her. We ſhifted our courſe to try to ſave her; we came up with her, put two men into her with oars, and a grappling, and then endeavoured to get on our way It might then be half an hour after eight. We ſtrove in vain againſt the tide, the violence of the waves and the wind. While we were putting the men and the oars into the ſmall boat, we had been driven to leeward more than three quarters of a league, on the ſide of the French iſland, ſituated near the coaſt, almoſt oppoſite the citadel. The darkneſs prevented us from ſeeing land, and indeed we could hardly diſcern the lights they had put out on board our two frigates

Perceiving therefore, that we got farther and farther from the ſhips inſtead of coming nearer them, we determined to make for land, and ſteered to the point where we thought the city was, for its ſituation was pointed out to us only by two lights, at a great diſtance from each other. The waves which broke againſt our boat, had already thrown in a great deal of water, which we emptied with our hats; we were wet to the ſkin, and the boat-men were much fatigued. M. de la Gyraudais, after having rowed for an hour, had now taken the helm; we knew not where we were, and had no brandy to keep up our ſtrength and ſpirits. In this diſtreſs we thought there was no better expedient for us, than to let fall our grappling, to give the men time to reſt themſelves. I then put on a great coat I found near me, and we diſtributed the quarter-cloths among the men, to cover themſelves with; not indeed to keep them from the waves, for we could not be more wet than we were, but to ſhelter them from the wind, which made us ſo very cold, that we were ob-obliged to ſqueeze as cloſe as poſſible to each other, in order to keep ourſelves warm. We were almoſt reſolved to remain in this condition all night, when M. de la Gyraüdais thought he perceived, that we were dragging our grappling. He ordered the ſteerſman to lay his hand on the hawſer, that he might judge by the motion, whether our grappling was really aweigh or not. The ſteerſman thought at firſt, that the motion he felt

was

was caufed by the fhocks the boat received from the waves; but he foon found out his miftake, and gave us notice of it. He was ordered to found with the boat-hook, which he did, and found only three feet water, with a bottom of rocks, which are on the borders of the whole coaft, and advance pretty forward in the river. The oars were placed in the row-locks, the grappling was drawn up, and we rowed for a full quarter of an hour, founding all the while, and finding the fame bottom. At laft we came to a muddy bottom, with feven or eight feet water. We were going to caft our grappling here, when the men forefeeing they fhould get no fupper in this place, faid, that as we were now in the way, we muft continue, and go and lay on fhore. We were extremely well fatisfied with their refolution, and fteered towards a light, which we imagined to be that of the guard placed at the only port where it is poffible to land.

Soon after, as we were all looking about us, endeavouring to find our fituation, we perceived a fchooner, which we knew to be at anchor very near the port. The fight of this veffel revived our fpirits, and we exerted ourfelves fo much, that in little more than half an hour we gained the port. The officer of the guard came out to reconnoitre us. Another officer was fent with the fteerfman to give the Governor notice of our being returned to the city, becaufe we had not been able to reach our veffel. He fent us compliments of condolance, and at the fame time invited us to fupper, and defired us to take up our night's lodging at his houfe.

We were apprehenfive of being troublefome to him, not only on account of the late hour, for it was midnight, but alfo becaufe we were too numerous a company: befides, as we were very wet, and in a ftrange pickle, we thought it better to go in fearch of a Frenchman named Lacombe, of St. Flour in Auvergne, fettled at Buenos Ayres, and having a houfe alfo at Montevideo: he was already known to many of our officers, from whom he had made feveral purchafes. A foldier of the guard, who fpoke French, offered to conduct us. Inftead of leading us to the place where M. Lacombe lived, he brought us to the houſe of a friend of his, where the ſoldier had ſeen him ſeveral times. We knocked near a quarter of an hour before we could get any anſwer. At length they anſwered, the door was opened, and we found M. de Belcourt in-bed in this houſe, as it was the place where he lodged. Thinking that we were playing him a trick, he did nothing but laugh at us. As we were not much in a humour for laughing, we enquired which was M. Lacombe's houſe, and were informed. We were making the beſt of our way to it when we met the Governor, who came on purpoſe to intreat us not to make uſe of any houſe but his. As we could not poſſibly refuſe, after many civilities on both ſides, we accompanied him.

When we came there we found every body up, and the cloth laid. Seeing we were all very wet, they offered us clean linen and cloaths. Thoſe who had put on their great coats ſooneſt, and were certainly not ſo wet as I was, refuſed even to change cloaths. The Governor's lady ſolicited me ſo often to put on at leaſt a night gown, that I at laſt accepted of it. It was one of her gowns, which I had ſo much trouble to get on that the Governor gave me one of his own. We ſat down to a light ſupper, provided haſtily for us. Our adventure was the ſubject of much converſation; at laſt we drank a diſh of chocolate, and, as it was now almoſt two o'clock, every body thought of retiring to bed.

Meſſ. de Bougainville and de Nerville were put into a ſmall room of the court yard, in which there was no other furniture beſides two chairs and two beds; one in a kind of alcove formed by a ſimple partition of wood, the other a camp-bed placed in the oppoſite angle. We had attended them to this apartment, and I took it for granted I was to paſs the night in a place much of the ſame kind, when a negro woman pulling me by the ſleeve beckoned me to follow her. She brought me back into the room where the company had been, where I found the Governor's lady and a negro woman employed in faſtening together

together ſome crimſon damaſk ſtools, which before ornamented the bottom of the room. Not knowing why ſhe was thus employed, I was talking in the mean time with the Governor. At length ſhe interrupted me, ſaying, it is for you, Sir, I am preparing this bed; you will be near us, and will not fare worſe than the reſt. After having expreſſed my gratitude for her attention and politeneſs, I did all I could to prevent her from going on with this buſineſs, but all to no purpoſe; ſhe ſtill proceeded, telling me it was a great pleaſure and ſatisfaction to her. When ſhe had made me a very good bed, ſhe retired with the Governor into the next room, where they ſlept.

A camp-bed was put up for M. de la Gyraudais, in the firſt room on coming in; and, as there were no more ſpare beds, a fire was made to dry our cloaths in the middle of the firſt hall, where M. de St. Simon and the reſt ſlept upon the chairs.

About half an hour after four, one of theſe gentlemen came to wake me, telling me M. de Bougainville was up, and that we muſt go. I dreſſed myſelf quickly, and we were going without ſaying a word, when a ſervant from the Governor came in, and deſired us to ſtay a little, that his maſter was getting up, and would pay his compliments to us. We told him to intreat the Governor from us not to riſe, that he wanted reſt, and that we were going away that inſtant, in order not to incommode him any longer. The wind and waves were conſiderably abated, and we got on board in a ſhort time.

The ſqualls of wind and rain our veſſels were expoſed to ſince the preceding evening had been extremely violent. The ſtorm having begun early in the ſpot where our frigates lay at anchor, becauſe they were not under ſhelter of the town as we were; the crews had been in very little concern about us, being perſuaded that we ſhould not even have run the riſk of coming from ſhore in ſuch weather. At all events, however, they had taken the precaution to put out lights. The two men we had put into the boat to ſave it, had been luckily driven into a ſmall ſandy creek under ſhelter of the French iſland; and the long-

boat

boat of the Sphinx, which had been ſent after her, had got to the bottom of the bay, in the place where we uſed to water. They were both returned when we came on board.

In the morning of the ſixth of January we went back to the town, to return our thanks to the Governor, and to pay him the compliments of the new year. He detained us to dinner. The converſation turned much upon the curioſities of the country. This made the Governor recollect that he had a ſhell which he thought very ſcarce. He ſhewed it us: it was a papyraceous Nautilus, as large and beautiful as any I ever ſaw. He made a preſent of it to M. de Bougainville. It had been ſent to him from Rio de Janeiro; and he told us, he had found a ſimilar one on the coaſt of the iſland Maldonnado; but that it had been broken. The Governor's lady gave me a parcel of the Canchalagua, which was all ſhe had left of it. A few days before ſhe had made M. de Bougainville a preſent of a paroquete, which ſpoke very prettily, and had alſo given him a cup made of the Calabaſh of Peru, mounted in ſilver, with a *Bombilla*, or tube of the ſame metal, uſed to ſuck up the maté.

Many authors of voyages have mentioned the Paraguay plant, or Caſſioberry buſh, as one of the principal ſources of the riches of the Spaniards, of the Indians, and eſpecially of the Jeſuits inhabiting this province. That my readers may be perfectly acquainted with this plant and its uſe, I ſhall inſert the account given of it by M. Ulloa, which he had from the miſſionaries of the country; for as they ſuffer none but their own brethren to penetrate into the country, this account can only be had from them.

" It is affirmed, ſays M. Ulloa, that the ſale of this plant was at firſt ſo conſiderable, and became ſo great a fund of riches, that luxury ſoon introduced itſelf among the conquerors of this country, who were at firſt reduced to the bare neceſſaries of life. As their taſte for luxury was always increaſing, in order to ſupport their prodigious expences, they were obliged to have recourſe to the Indians ſubdued by force of arms, or who had voluntarily ſurrendered themſelves: of theſe they made their ſervants, and ſoon after their ſlaves. They worked them too hard, ſo that many of them fell under the weight of labour they were unuſed to; and more of them under the oppreſſion of the cruel treatment they were expoſed to, rather from the loſs of their ſtrength, than from their indolence. Others eſcaped by flight, and became moſt irreconcileable enemies to the Spaniards. The Spaniards fell into their former ſtate of indigence; which however did not make them more induſtrious. Luxury had increaſed their wants ſo much, that the ſale of the Paraguay plant alone was not ſufficient to ſupply them: moſt of them indeed were now no longer able to buy it, for the great conſumption of it had enhanced its price." Tom. I. page 13.

This plant, ſo famous in South America, is the leaf of a tree about the bigneſs of a middling ſized apple-tree. Its taſte is like that of the mallow, and in figure it nearly reſembles the orange leaf. It alſo bears ſome reſemblance to the leaf of the *Cocoa* of Peru, where a great deal of it is carried, eſpecially among the mountains, and in all places where they work the mines. The Spaniards think it the more neceſſary, as the uſe of the wines of the country is hurtful there. It is brought dry, and almoſt reduced to powder. It is never ſuffered to remain long in infuſion, becauſe it would then turn the water as black as ink

It is diſtinguiſhed into two kinds, though they are both one and the ſame leaf. The firſt is called *Caa*, or *Caamini*; the other *Caacuys*, or *Yerva de Palos*; but Father *del Techo* aſſerts that the name of the genus is *Caa*; and diſtinguiſhes three ſpecies, under the names *Caacuis*, *Caamini*, and *Caaguazu*.

According to the ſame traveller, who paſſed great part of his life in the Paraguay, the Caacuys is the firſt bud, juſt beginning to expand its leaves. The Caamini is the leaf in full growth, from which the ſtalks are taken, before it is roaſted: if the ſtalks are left on, it is called Caaguazu or Palos. The leaves when roaſted are preſerved in pits digged in the earth, and covered with

with a cow's ſkin. The Caacuys will not keep ſo long as the two other ſpecies, the leaves of which are exported to Tucuman, to Peru, and even into Spain. It is very liable to injury in the removal. It is aſſerted even, that this plant, taken on the ſpot, has a particular bitterneſs, which enhances its virtues as well as its price, and which it loſes by tranſportation.

The manner of taking the Caacuys is by filling a veſſel with boiling water, into which the leaf, powdered and reduced to a paſte, is thrown. As it diſſolves, any ſmall portion of earth which may have remained ſticking to it, floats at the top, ſo as eaſily to be ſkimmed off. The water is then ſtrained through a rag, and being ſuffered to ſtand a little, is afterwards ſucked up through a reed. Generally there is no ſugar put to it; but a little lemon juice is mixed with it, or ſome kinds of wafers of an agreeable ſmell. When it is taken as a vomit, a little more water is thrown on it, and it is left till it is almoſt cold.

The moſt famous place for this plant is at *Villa*, or the new *Villarica*, which is near the mountains of Maracagu, ſituated to the Eaſt of Paraguay; about 25 degrees 25 minutes South latitude. This diſtrict is celebrated for the cultivation of this tree; it is not however upon the mountains that it grows, but in the marſhy bottoms which divide them.

From this place are taken for the Peru only, one hundred thouſand *Arrobes*, each of which weighs twenty-five pounds of ſixteen ounces ſtandard weight, and the price of the arrobe is ſeven crowns, or twenty-eight French livres, ſo that the whole profit of the hundred thouſand Arrobes amounts to two million eight hundred thouſand French livres. Yet the Caacuys bears no fixed price, and the Caamini ſells for twice as much as the Caaguazu. The laſt of theſe, while we were at anchor at Montevideo ſold for twenty-five livres, or five piaſters per Arrobe. The Governor procured it us at this price.

The Indians ſettled in the provinces of Uraguay and Parana, under the government of the Jeſuits, have ſown ſome of the ſeeds of this tree, brought from Maracayu, which have hardly degenerated

degenerated in the leaſt. They reſemble much the ſeed of the ivy. But theſe Indians do not prepare the plant in its firſt ſtate; they keep the Caamini for their own uſe, and ſell the Caaguazu or palos to pay the tribute they owe to Spain.

The Spaniards imagine they find in this plant a remedy or preſervative againſt all diſeaſes. Every body agrees that it has a laxative and diuretic quality, but I would not anſwer for all the properties the Jeſuits attribute to it. I believe the moſt incontestible of theſe properties, which is however the one they are moſt ſilent upon, is that of ſupplying them with a prodigious ſum of money every year.

It is reported, that ſome perſons having at firſt taken this plant too freely, it brought on a total deprivation of their ſenſes, which they did not recover till a few days after. It appears indeed certain, that it often produces oppoſite effects, ſuch as to procure ſleep to thoſe who want it, and to rouze thoſe who are lethargic; to be at once both nouriſhing and purgative.

Cuſtom renders the uſe of it neceſſary, and it is often with difficulty that people abſtain from an immoderate uſe of it; for it is affirmed that an over-doze of it inebriates, and brings on moſt of the inconveniencies which follow an exceſs in drinking ſtrong liquors.

According to Mr. Ulloa, the Paraguay plant, is called *Maté* at Peru. He ſays, that in order to prepare it a certain quantiy is thrown into a calabaſh, mounted in ſilver, which is alſo called *Maté*, or *Totumo*, or *Calabacito.*

Some ſugar is thrown into this veſſel, and cold water poured upon the whole, that the plant reduced to a paſte may be well moiſtened: the veſſel is afterwards filled with boiling water; and the plant being in very ſmall pieces, the liquor is ſucked up through a tube of a ſufficient ſize, but too ſmall to admit the plant to paſs. The tube or reed made uſe of is called *Bombilla.* As the water diminiſhes it is renewed, adding always ſome ſugar, till the plant floats no longer on the ſurface: at which time a fresh quantity of it is put in. It is often mixed with the juice of lemon, or Seville orange, and with sweet scented flowers. This liquor is commonly taken fasting: but many people drink it also after dinner. Perhaps the plant may be heathful; but the method of taking it is extremely disgusting. How numerous soever the company is, every person drinks by turns through the same tube or bombilla, handing the *maté* from one to the other. The Spaniards of Europe care very little for this drink, but the Creoles are passionately fond of it. They never travel without a supply of the Paraguay plant; and never omit taking it every day, preferring it to all other kind of food, and never eating any till after they have taken it.

Some, says Frézier, (Relat. du Voyage de la Mer de Sud, page 228) call the Paraguay plant, St. Bartholomew's plant; because they say this Apostle went into these provinces, where he changed the plant which was before poisonous, and made it salutary and wholesome. Instead of drinking the infused liquor separately as we do tea, they put the plant into a cup made of a calabash, mounted in silver, which they call *Maté*. They put sugar to it, and pour hot water over it, then drink it off immediately, without suffering it to stand in infusion, because it would grow as black as ink. In order to avoid taking up the plant, which floats on the surface, they make use of a silver tube, the end of which is formed into a round knob, perforated with several small holes. so that the liquor, which is sucked through the other end, comes up without the plant. The company drinks round through the same tube, pouring more hot water as the first is consumed. Instead of the reed or *bombilla*, some people remove the plant with a plate of silver, full of small holes. The aversion which the French have shewn to drink after all kinds of people, especially in a country where there are so many persons afflicted with the venereal disease, has introduced an invention of small glass tubes, which they now begin to use at Lima. In my opinion, the taste of this liquor is better than tea, it has an agreeable vegetable smell. The

people

people of the country are ſo uſed to it, that even the pooreſt among them drink it at leaſt once a day.

The trade of the Paraguay plant, adds the author, is carried on at *Santa Fé*, where it is brought by the river Plata, and by land carriage. There are two kinds of it: one of which is called *Yerva de Palos*; the other, which is a finer ſort, and of a higher quality, is called *Hierba de Camini*. The laſt comes from the lands belonging to the Jeſuits. The greateſt conſumption of it is made from *Paz* to *Cuſco*, where it is twice the value of the other, which is ſold from Potoſi to Paz. Above 50,000 arrobes are brought every year from Paraguay to Peru; that is to ſay, 1,250,000 weight of one and the other ſpecies, the third part of which quantity at leaſt is Camini; without reckoning about 25,000 Arrobes of the Palos, ſent into Chili. Each packet, containing ſix or ſeven Arrobes, pays four reals duty, at Alcavala; and the expence of conveying it above ſix hundred leagues doubles the prime coſt, which is about two piaſters: ſo that at Potoſi, it coſts five piaſters, or five and twenty livres of France, per Arrobe. It is generally conveyed in carts, which carry one hundred and fifty Arrobes from Santa Fé to *Jujui*, the laſt town of Tucuman; and from thence to Potoſi, which is ſtill an hundred leagues farther, it is carried upon mules. I have obſerved, that the uſe of this plant is neceſſary in places where they work the mines, and in the mountains of Peru, where the white people imagine wine to be pernicious: they rather chuſe to drink nothing but brandy, leaving the wine to the Indians and Blacks, who are very well ſatisfied with it.

I was witneſs at Montevideo of the truth of the account given by theſe two authors. At whatever time of the day one goes into any houſe, one is ſure to find ſomebody drinking *Maté*, which they never fail to offer to any one who comes in, even in the very hotteſt weather; being perſuaded that this infuſion is cooling, that it aſſiſts digeſtion, &c. The veſſel out of which one drinks the *Mate* uſually ſtands on a foot, faſtened to a board. This was the general cuſtom in almoſt every houſe; but ſome

X 2 of

of the inhabitants held the veſſel alone, ornamented with ſilver, in their hands, without any board. There are alſo ſome *bombillas* or reeds, the end of which put into the liquor, is formed like an oyſter-ſhell, fixed to the tube by the top of its hinge.

While we were at the Governor's, two of our ſailors deſerted; ſome ſaid that one of them was a Malteſe, others that he came from Biſcay: they had given him the nick-name of Spaniard. The other came from lower Britanny. We ſearched for them in vain. We learnt afterwards that they had offered to enter on board the Spaniſh frigate, St. Barbe, but the owner aſſured me he had refuſed them. A few days after four ſailors deſerted from the Sphinx: one of them named Plaiſance, who had been formerly a dragoon, and had ſerved in Canada under M. de Bougainville. This man had been very preſſing to let him embark with us, when we left St. Malo. M. de Bougainville had always looked upon him as a very honeſt and brave man, very fit to become an inhabitant of a colony. He had given him two complete ſuits, and other cloaths. Two days before he had deſerted, he had been truſted with a fuſil, and a rich ſword to ſell. He gave out that they had been ſtolen from him. Whether this report was true, or whether he had really ſold them, it is certain that one of the inhabitants declared that Plaiſance had ſold the ſword to an officer's ſervant. Plaiſance finding himſelf ſuſpected, and not being able to clear himſelf properly of the accuſation, ran away, for fear of being puniſhed for his diſhoneſty. The Governor, at M. de Bougainville's ſolicitation, who had promiſed ten piaſters for every deſerter that ſhould be brought back to him, ſent ſome dragoons after them, but they came back without any tidings of them. I believe that if one had even promiſed a reward of one hundred piaſters, they would not have ſtopped any of them: for it is the intereſt of Spain to retain as many men as poſſible in the country for the ſake of population.

Monte-

A VIEW of the TOWN of St. PHILIP of MONTE VIDEO.
Monte Video
Citadel
the Parish Church
the Gate
Frenchmans Island
the Jetty
Franciscan Fryars
Windmill

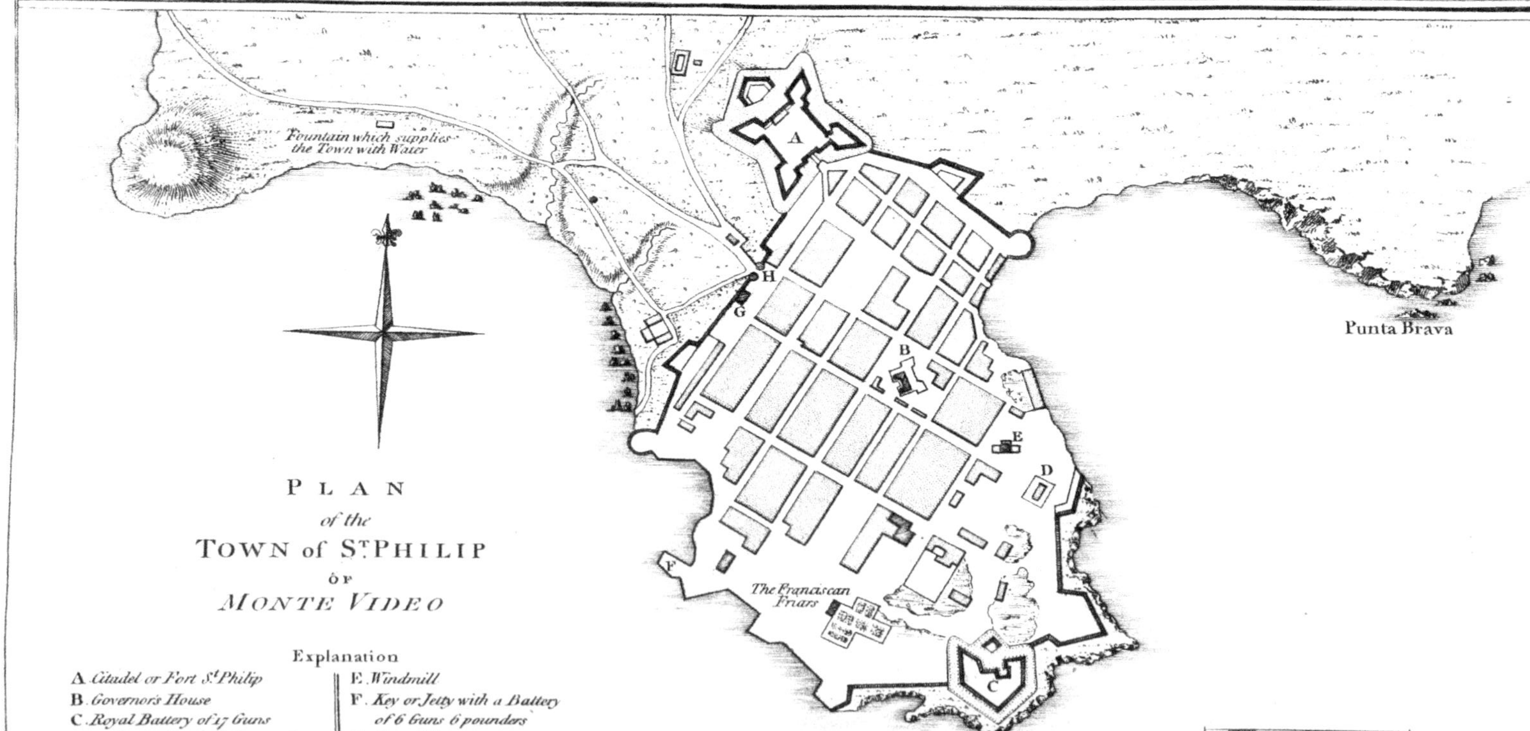
PLAN
of the
TOWN of St. PHILIP
or
MONTE VIDEO
Explanation
A. Citadel or Fort St. Philip
B. Governor's House
C. Royal Battery of 17 Guns
E. Windmill
F. Key or Jetty with a Battery of 6 Guns 6 pounders
G. Guard Room
Fountain which supplies the Town with Water
The Franciscan Friars
Punta Brava
A
B
C
D
E
F
G
H

Montevideo is a new colony. Five and twenty years ago there were only a few huts in it. It is, however, the only tolerable place for the anchorage of ſhips that come up the Rio de la Plata. At preſent it is a ſmall town which improves every day. The ſtreets are made perfectly ſtrait, and wide enough to admit three coaches abreaſt. I have given a view of it, taken as it appeared from on board the Eagle frigate, while we lay at anchor between the mountain and the town, according to the deſcription I have given above.

The houſes conſiſt only of ground floors, under the roof; one of them only is to be excepted; this is ſituated in the great ſquare, and belongs to the engineer, who built it and lives in it. It has one ſtory, and a kind of a double roof with a pretty long projection, which ſupports a balcony in the middle of the front. I have given the plan of this town.

Each tradeſman's houſe generally conſiſts of a hall, which ſerves by way of entrance, a few bed-chambers, and a kitchen, the only place in which there is a chimney, and where they make any fire. Theſe houſes are therefore properly a ground floor, fourteen or fifteen feet high, including the roof. The entrance into the Governor's houſe is a long ſquare hall, which receives light only from one ſmall ſaſh, half filled with paper and half with glaſs. The bottom of the ſaſh is cloſed with planks of poliſhed wood. This hall may, perhaps, be about fifteen feet wide, by eighteen feet long. From hence one goes into the room for receiving company, which is almoſt ſquare, but rather longer than it is wide. At the bottom, oppoſite the only window in the room, made much in the ſame taſte as that I have before deſcribed, there is a kind of alcove ſix feet wide, cloſed with bars of iron, and covered with tigers ſkins. In the middle of this is an arm chair for the Governor's Lady, and on each ſide ſix ſtools covered, as the arm chair is, with crimſon velvet. All the ornament conſiſts in three ſmall bad pictures, and a few large plans, half-pencilled and half coloured, ſtill worſe in point of drawing than the pictures. The two other

ſides

ſides of the room are filled with ſeats for the men. Theſe are wooden chairs with very high backs, reſembling thoſe made in the time of Henry the fourth of France, having two turned pillars ſupporting a frame which adorns the middle, covered with leather, curiouſly ſtamped and wrought, as well as the ſeat. The door which leads from this room into the next, where the Governor and his lady ſleep, is only cloſed by a kind of curtain made of tapeſtry. The two angles of this room on each ſide of the window are filled up, one with a wooden table, upon which the veſſel for taking the *Maté* always ſtands; the other with a kind of cupboard, having two or three ſhelves, furniſhed with a few china diſhes and cups.

The lady of the houſe is the only perſon who ſits in the alcove when there are only men in company, except ſhe ſhould invite ſome of them to ſit on the ſtools near her.

Theſe rooms have, generally ſpeaking, neither flooring, nor pavement. From the inſide of them one may ſee the reeds which ſupport the tiling of the roof.

The white people ſpend their time in idle converſation, in taking the *Maté*, or in ſmoaking a *Sigare* or *Cigare*, which is a kind of ſmall cylinder, ſix or ſeven inches long, and about half an inch in diameter, compoſed of tobacco leaves rolled one over the other.

The merchants, and a very few artiſts, are the only perſons who have any employment at Montevideo. There are no ſhops, no ſigns, nor no outward ſhow, by which they can be found. But one is ſure of meeting with them, if one goes into any houſe ſituated in an angle formed by the meeting of two ſtreets. The ſame merchant ſells wine, brandy, woollen drapery, linen, toys, &c.

In the ſtreets one meets with nothing but white or black people, or mulatoes on horſeback; and horſes ſtanding at the doors of the houſes without being faſtened. This country might well b called a hell for horſes. They often make them work three days following, without giving them either meat or drink; ſometimes

ſometimes they are kept tied up for as long a time, with the ſame treatment, and doing nothing except running from the end of one ſtreet to another. At the end of three days, they are ſent back again into the country to feed upon what graſs they can find. The perſon who goes with them, takes off the ſaddle, and puts it upon another horſe, whom he brings to the town to be treated in the ſame manner.

Notwithſtanding this they are excellent cattle, having preſerved the ſpirit of the Spaniſh horſes, from which they are bred. They are extremely ſure footed, and ſurprizingly ſwift. Their ſtep is ſo ſharp, and ſo long, that it is equal to the full trot, or ſmall gallop of our horſes. Some of them are ſo light that nothing can be compared to them. When they ſtep they raiſe the fore foot and the hind foot at the ſame time; and inſtead of bringing the hind foot in the place where the fore foot was, they ſtretch it out much farther, bringing it oppoſite to and even beyond the fore foot of the other ſide; which makes their motion as quick again as that of other horſes, and at the ſame time much eaſier for the rider. They are not remarkable for beauty; but deſerve much encomium for their ſwiftneſs, mildneſs, courage, and abſtemiouſneſs. The inhabitants make no proviſion of hay or ſtraw for theſe animals. Their only food all the year is in the fields. It is true, that in this country it is never cold enough to freeze either the rivers or the plants.

The environs of Montevideo are an extenſive plain. The ſoil is a black thick earth, extremely fertile with very little manurement. This country only wants ſome perſons to be employed in cultivating it to become one of the beſt in the world. The air of it is wholeſome, the ſky ſerene, and the heat not exceſſive. It is rather deficient in wood, which is found only a long ſide the rivers. Here tigers, leopards, and other wild beaſts chiefly reſort. The tigers eſpecially are rather numerous, larger, and more fierce than thoſe of Africa. The Governor had one of theſe tigers brought up from a whelp in his court yard. He was faſtened near the entrance of the door, with a ſingle

ſingle ſtrap of leather, paſſed round his neck. The dragoons and ſervants uſed to play with him, and he never gave any ſigns of his natural ferocity. They uſed to turn him about, to pull him, to throw him over and over as one would do a tame cat. The Governor ſeeing that M. de Bougainville took a fancy to him, had him carried on board, and made a preſent of him. A cage was made for him of thick planks, ſix inches in ſquare, and he was kept eight days At the end of this time he began to roar now and then, eſpecially in the night. It was then apprehended that he would grow furious, or that, even in play, he might ſwallow the arm of ſome of the ſhip boys, or children who went to ſee him, and who ſometimes put their hands between the planks of his cage. Beſides, it was neceſſary to ſupply him with freſh meat for his food, and we had none of that to ſpare. Theſe conſiderations determined M. de Bougainville to have him ſtrangled. He was then but four months old, and his height, when he ſtood upright, was two feet three inches. By this one may judge how high he would have grown.

The Spaniards of Montevideo live, as I have ſaid before, in great indolence. They are cloathed nearly as the Portugueſe at the iſland of St. Catherine; but they very frequently wear white hats, the flaps of which hang looſe over their ſhoulders, and cannot be made too large for them.

The women are pretty well ſhaped, but one cannot ſay with tru h that they have a complexion of lilies and roſes; on the contrary they are much tanned, have commonly but few teeth, and thoſe not white.

Their dreſs conſiſts outwardly of a plain white or coloured waiſtcoat, well fitted to the waiſt, the ſkirts of which fall four fingers in length upon the petticoat. This petticoat is made of a ſtuff more or leſs rich, according to the circumſtances or fancy of the perſon who wears it. It is edged with gold lace, or with a fringe of ſilver, gold, or ſilk; ſometimes in double rows, but without flounces. They wear no caps of linen or lace. A ſimple ribbon paſſed round the head keeps the hair together at the

A Spaniard of Montevideo.

A Spanish Lady of Montevideo.

top, from whence paſſing on the back part of the head, it falls in two or three treſſes down the back; flowing ſometimes as low as the bend of the knee. The longer they are the more beautiful they are reckoned.

When the women go out, and ſometimes even in the houſe, they cover their heads with a piece of fine, white, woollen ſtuff, trimmed with gold or ſilver lace, or ſilk. This piece of ſtuff which they call *Iquella*, or mantle, covers alſo their ſhoulders and arms, and falls down below the waiſt. They croſs the ends of it over the breaſt, or under the arms, as our French ladies do their cloaks. When they wear this kind of mantle in the houſe, they ſeldom cover their head with it. The country women of Poitou wear ſome nearly of the ſame kind. But in the ſtreets, and at church, the Spaniſh women put this mantle ſo cloſe upon their heads, that one can hardly ſee any of their face except an eye, and the noſe: in the houſe they often do not even cover their neck with it.

The women at home enjoy at leaſt as much liberty as in France. They receive their company with much politeneſs, and are eaſily prevailed upon to ſing, dance, play upon the harp, guitar, theorbo, or mandoline. In theſe things they are much more complaiſant than our French ladies. When they are not engaged in dancing, they ſeat themſelves upon ſtools raiſed, as I ſaid before, under a kind of alcove, at the bottom of the room, where the company is. The men cannot ſit near them, unleſs they are invited; and when this favour is beſtowed upon them, it is looked upon as a mark of familiarity.

The manner of dancing among the ladies ſeems to partake of the indolence in which they paſs their lives, though they are naturally very lively. In moſt of their dances their arms either hang looſely down by their ſide, or are folded under their mantle, which they alſo call *Rebos*. In going through the *Sapateo*, one of the moſt common of their dances, they keep their arms raiſed up, and ſnap their fingers in the air, as they ſometimes do in France, when they dance the rigadoon. The Sapateo is performed without moving much out of the ſame place, and by ſtriking the ground with the extremity of the foot and the heel alternately. The ladies hardly appear to move; they rather ſeem to ſlide along upon their feet, than to advance in cadence; this is owing to the lightneſs and celerity with which they move their feet.

The Governor and the military men are dreſſed after the French faſhion, except that they always wear a hat upon their heads, and that they are never powdered or curled, any more than the women. They live alſo in a great ſtate of indolence, as well as the other Spaniards, who are dreſſed nearly in the ſame manner as the Portugueſe of St. Catherine's iſland.

The common people, mulatoes, and negroes, inſtead of a cloak, wear a piece of broad ſtriped ſtuff, of different colours, ſlit only in the middle, to let the head through. It falls down upon their arms, and covers them as low as the wriſt. On the fore, and on the back part, it comes down below the calf of the leg, and is fringed all round. This garment is called *Poncho*, or *Chony*. Every body wears it on horſeback, finding it more convenient than the cloak or great coat. The Governor ſhewed us one of them, wrought in gold and ſilver at Chili, from whence this garment has been brought. It had coſt him more than three hundred piaſters: ſome of them are made at the ſame place at the rate of two thouſand.

The Poncho keeps off the rain, and defends from the wind; it ſerves for a bed covering at night, and for a carpet in the country. All theſe dreſſes may be ſeen in the plates.

The Spaniards live in a very plain manner. The men, who are not buſied in trade, riſe very late, as well as the women. The ſlaves, negro women, or mulatoes, prepare the *Maté*, while their maſters are dreſſing, who put the reed into their mouths, almoſt before they have put their feet into their ſlippers. The men afterwards ſit ſtill with their arms folded, till they take it in their heads to converſe, and ſmoke a *cigale* with their neighbours. Four or five of them ſometimes ſtand together

A Spaniard of Montevideo.

A Spanish Gentleman of Montevideo.

ther at the door of a houſe, talking and ſmoking. Others mount their horſes, and go out, not to take a ride about the country, but through the ſtreet. If they have a fancy for it, they get off their horſes, mix with any company they meet, goſſip for two hours without ſaying any thing of conſequence, ſmoke, take ſome *Maté*, then mount their horſes again; who has been all this while ſtanding as ſtock ſtill as a wooden horſe, without being faſtened, and as if he was liſtning to the converſation. Sometimes there are as many horſes as men.

During this interval, the women remain ſeated on a ſtool at the bottom of their apartment; having under their feet next to the floor a mat made of reeds, and over the mat ſome cloaks of the ſavages, or ſkins of tigers. There they play upon the guitar, or upon any other inſtrument, which they accompany with the voice; or they take the *Maté*, while the negro women are dreſſing the dinner in the ſame room.

About half an hour after twelve, or one o'clock, the dinner is ſerved up; this conſiſts of beef dreſſed in various ways, but always with a great quantity of pimento and ſefran. Sometimes ragoos of mutton are brought up, which they call *Carnero*; and ſometimes fiſh, but very ſeldom any poultry, which is rather ſcarce. There is great plenty of game, but the Spaniards do not go in queſt of it, as that would be too fatiguing. The deſert is compoſed of ſweetmeats.

Immediately after dinner, both maſters and ſlaves indulge in the *Sieſta*, that is, they lie down; ſometimes they undreſs themſelves and go to bed, where they ſleep for two or three hours. Workmen, who live by the labour of their hands, do not deny themſelves theſe hours of indulgence. A great part of the day being thus loſt, this is the reaſon of their doing but little work. and makes all handicrafts exceſſively dear. This circumſtance may alſo proceed from the plenty of money there is here.

It is not ſurprizing they ſhould be indolent and lazy. Their meat coſts them only the trouble of killing, ſkinning, and cutting up the ox to drefs it. Bread is very cheap. The fkins of oxen and cows ferve to make them all kinds of facks, to cover part of their houfes, and for a thoufand other purpofes, for which different forts of materials are ufed in Europe. Thefe fkins are fo common, that many flips of them are found fcattered here and there along the ftreets the leaft frequented, in the fquares, and upon the walls of the gardens.

Few of thefe gardens are cultivated, though there is one belonging to each houfe. The ground is left fallow. I faw but one garden tolerably well kept, and this undoubtedly was becaufe the gardener was an Englifhman. Vegetables therefore are fcarce here. The plant they cultivate the moft is the fefran or Carthamum, for their foup and fauces.

It is very common among them to keep a miftrefs. Thofe who have children by them, give thefe children a kind of legitimacy, by acknowledging themfelves publickly to be their fathers: after which, thefe children inherit nearly as the legitimates do There is no ignominy fixed upon illegitimate births; becaufe the laws authorize them fo far, as even to beftow the title of gentlemen to baftards: in which thefe laws appear more agreeable to humanity, not making the innocent fuffer for the guilty.

I have obferved, while I was at mafs, that the Chaffuble was compofed only of three flips of ftuff, fewed together lengthways, without being in the form of a crofs. The middle flip only is of a different colour from the other two. During the time of mafs, one of the inhabitants plays upon the harp, in a gallery: this harp certainly ferves inftead of an organ. I faw no particular demonftration of devotion, but that of ftriking their breafts pretty hard five or fix different times, from the beginning of the fervice till after the communion. The Rofary is much in ufe here; and the Ave Maria is almoft the only prayer they fay. Many of them wear the Rofary round their necks. The Portuguefe of St. Catherine s ifland, white men, negroes, and mulatoes, had likewife almoft all of them Rofaries; fome wore them

them outwardly, eſpecially the blacks, the others wore them under their cloaths. They are alſo very devoutly inclined to the ſcapulary of mount Carmel; which is worn by both men and women. They think the ſcapulary and the *Avillas* will preſerve them from all dangers, and inſure their eternal ſalvation. They are ſcrupulous only about the externals of devotion. Theſe Avillas which one ſees hanging at their necks are a kind of ſea cheſnut, reſembling a flat round bean, of the ſize of a half-crown, and two lines and a half in thickneſs; the ſkin is granated, and very finely ſhagreened, of a pale cheſnut colour; at its circumference there is a black band, which almoſt ſurrounds it. I picked up a great many of them on the ſea coaſt, at the iſland of St. Catherine, without knowing what they were: and I have ſeen many of them mounted in ſilver, at a goldſmith's ſhop in Montevideo. He told me, that when it was worn round the neck it preſerved from infectious air and witches.

At each altar there is a veil which reaches from top to bottom, always hanging before the principal image, in the ſame manner as that they place in France before the hoſt, when taken out of the tabernacle, during the time of a ſermon or a diſcourſe. This veil always remains. At the beginning of maſs, the ſtring which faſtens the veil is pulled up, and the veil raiſed like a curtain, ſo as to diſcover the image: when maſs is over, the veil is let down again.

Two days after ſailing from Cape Frehel, near St. Malo, we put into a ſmall barrel of water a liquor which had been given us by M. Seguin, a chymiſt, who lives in the *rue des poſtes, prés de l'Eſtrapade*, at Paris, as a preſervative of ſweet water from corruption, as well by ſea as by land, and as having the property not only of preventing, but likewiſe of curing the ſea ſcurvy. As we had hitherto kept the water we brought with us from St. Malo's, ſweet; we now compared it with the other, and finding no difference, we determined not to open this caſk again, in which we had put the liquor, till the freſh water ſhould undergo ſome conſiderable change.

8 The

The very night in which the ſtorm obliged us to lie at the Governor's, it produced more fatal conſequences, at the diſtance of two cannon ſhot from our frigates. The thunder was very loud, and the lightning fell upon the Spaniſh veſſel the St. Barbe, which had ſhifted her anchorage two days before, in order to be more at hand for ſailing out of the river with the firſt favourable wind. Their change of poſition brought on this accident, by which they had one man killed, and fourteen wounded, five of whom were dangerouſly hurt; and their mizen maſt was ſhattered.

The next day we carried to the Governor's houſe, the compaſs invented by captain Mandillo a Genoeſe, for finding out the longitude. We wanted to make ſome obſervations upon land, which we had not been able to do upon the veſſel all the time of our voyage, even when it was calm; becauſe the fault of this compaſs, is, that the leaſt motion diſturbs the ſteadineſs of the needle. During a calm, even when it lies quite even, it is more or leſs agitated. Notwithſtanding all poſſible care was taken to preſerve this compaſs, yet the damp of the ſea air, which penetrates every where, had affected the needles, ſo as to make them a little ruſty at the center, and near the parts which keep them in equilibrio. They had therefore loſt that property which is neceſſary, and their magnetic virtue was likewiſe ſomewhat impaired. We cleared them from the ruſt, and recovered their magnetic powers; but we remitted our obſervations till another day, becauſe it was now late, and we left the inſtrument with the Governor.

On this occaſion we expreſſed our aſtoniſhment to the Governor, that the inhabitants of Montevideo ſhould not think of procuring themſelves ſhade in their gardens, and other extenſive places, by planting of trees; and we mentioned our ſurprize alſo at obſerving, that the country ſeemed totally deprived of that benefit. He told us there were ſome trees along ſide the rivers, and that a country houſe which he had at the diſtance of about two leagues from the city was well furniſhed with them.

them. He propoſed a party to go there on horſeback the next day in the afternoon, after dining with him. We accepted of the ride with an intent not only of ſeeing the country, but at the ſame time to convince ourſelves of the many wonderful and incredible things he and many others had told us of the horſes of Paraguay.

The party being concluded on, the Governor took upon himſelf to provide us with a ſufficient number of horſes, which were to be ready about three or four o'clock.

The vicar had invited me to dine with him that day, in company with Meſſ. Duclos, the two brothers, firſt and ſecond-captain, M. de Belcourt, the paymaſter of the Spaniſh troops, a Fleming who ſpoke the French language well, and the two ſons of M. Duclos the elder. We went there, and during the whole dinner time, a mulatto played upon the harp. About the middle of dinner, another man, whom they called a civilized Indian, joined the former and accompanied him with his guitar. Then the vicar, who was the only eccleſiaſtic in town, called in four or five little blacks about eight or ten years old, and as many negro girls of the ſame age. He made them dance to the ſound of his inſtruments, and the caſtanets they had in their hands. The children acquitted themſelves with ſurpriſing agility and cleverneſs. One circumſtance a little tireſome in theſe Indian dances, is, that almoſt all the ſame motions are repeated in every dance. It muſt alſo be allowed that there is no great variety in the tunes they ſing or play. The muſick of ſome of them, eſpecially of the Sapateo is pricked down in Frezier's account of a voyage to the South Sea.

They knew in this country, not only what the King of Portugal had done againſt the Jeſuits in his dominions, but alſo what the parliaments of France, and the government, had enacted againſt this ſociety. The vicar deſired me to give him in writing an account of what was repreſented in that famous picture found among the Jeſuits at Billom, in Auvergne, at the time when an inventory was there taken of the furniture and

effects

effects of those fathers, after the condemnation and suppression of their institution in 1762, and 1763; and the secularization of its members. I satisfied his curiosity with regard to this authentic monument of Jesuitical folly. This vicar is a man of good sense, and generally beloved. He has thirty slaves, negroes and negro-women, great and small. It is his pleasure always to have somebody to dine with him. He gives his company a hearty welcome, and treats them well. He looks upon all his slaves as his children, and is beloved by them. He educates them well in order to give them afterwards their liberty, with forty or fifty cows or oxen, to put them in a way of keeping up their freedom. But he pays a particular regard, I may indeed say, he shews an uncommon degree of affection, for a little mulatto, almost white, who is the son of one of his mulatto slaves, as he was pleased to say by an Irish officer, but who bears in all the features of his face the strongest marks of being the vicar's own son. He told us he intended to send him into France to study, and to make a physician of him. The child is at present seven years old. When there is company, he makes him dine by himself, and when there is nobody, often with him. He has already settled five and twenty thousand piasters upon him. His living, and his own private income bring him in about four thousand, and he is now about sixty years old.

We were waited on at table by four negro women, by the mother of the little mulatto, who is also a mulatto, and by an Indian woman the wife of a Cacique, taken at the colony of St. Sacrament from the Portuguese, in the last siege the Spaniards laid to it. These women were all with child, though neither of them was married except the Indian, who knew not whether her husband was alive or dead. Neither men nor women in this country, are in the least scrupulous upon this article.

At the desert, Mess. de Bougainville, de Nerville, and l'Huillier de la Serre, came to meet us; and we all went together to the Governor's house, where we found horses ready for us. The

Governor's

Governor's lady, dreſſed like an Amazon, and having a gold laced hat on, cocked after the military faſhion, put herſelf at the head of our cavalcade, mounted upon a beautiful horſe, whoſe value anſwered to his appearance. M. de Bougainville's horſe was equally fine. With a kind of doubled pace, which reſembled ambling, they always left us behind. It was all we could do to keep with them, ſome of us trotting, others on a canter. We continued this pace till we reached the country houſe, which we did not do till more than a full hour after our firſt ſetting out, though they told us it was no more than one league diſtant from Montevideo.

Father Rock, a Franciſcan, was here waiting for us, with the Governor's ſon, a child of three years and a half, to whom this father was preceptor. We found a plentiful collation ready for us; and after drinking a glaſs, merely on account of the heat of the weather, we went to ſee the Governor's orchard, which he called a wood.

This country ſeat is nothing extraordinary with regard to the houſe; which conſiſts of one ſingle ground floor, as all the other houſes do; on account of the violent winds, very frequent in this country, which might blow them down if they were built higher. The only remarkable thing here is a tolerably pretty room, which however has no other ornament than ſome geographical maps, fixed on the bare wall, and ſome wooden chairs covered with leather, which is figured with flower work.

At the diſtance of two or three gun ſhots from the houſe, the orchard is planted: this conſiſts of apple, pear, peach, and fig trees, diſpoſed in alleys, which are not very regular, except the middle walk, which reaches from one end of the orchard to the other, and is about a mile and a half in length. A pretty conſiderable brook winds through the orchard; which has probably hindered the walks from being cut ſtrait. They are however extremely rural, on account of the number of tall, as well as low plants growing in them without cultivation. The balm eſpecially abounds there. I acquainted the Governor, M. Belia, and another officer, with its virtues. They were the more pleafed with my account, as the plant is extremely plentiful in this country, and they thought it might perhaps be ufed inftead of the *Maté*.

The trees were fo laden with fruit, that moft of the branches, unable to fupport the weight, were already broken. We advifed the Governor to have the others fupported with props, more efpecially as he told us, that all thefe fruits were of the beft and moft excellent kind. We could not judge of this ourfelves, as the time of their being ripe, was not till the end of February; at prefent however they had an exceeding fine appearance.

One might make a delightful walk of this orchard; but the Governor does not employ any body about it, becaufe it is his intention to return to Europe, where he propofes to fix entirely.

As we were coming back, I fpoke to father Roch, and held a converfation with him in Latin, upon feveral points of philofophy; which I foon found he had only ftudied in Ariftotle's fchool, both by the barbarous and obfolete terms he made ufe of, and by the fyftem he followed. He indeed confeffed himfelf to be much attached to it. He told me, he was *a Peripatetic and a Scotift, and would remain fo all his life.* He fpoke pretty good Latin, and with facility. The greateft difficulty to me was his pronunciation of u as ou, and his manner of pronouncing the g, which the Spaniards always lofe in their throats, pronouncing it nearly as an afpirate. Befides the attention this required of me in order to underftand him, I was alfo obliged to confider of what I had to fay, and at the fame time to endeavour to catch his pronunciation, without which, he might not perhaps have comprehended me. A few days before, for the firft time, I had been exactly in this fituation with him. Having heard that he was a man of learning, I had been to pay him a vifit at his convent. I inquired for him, in the Latin tongue, of one of his brethren who opened me the door. He made me a fign to walk in, without anfwering a fingle word. I went in, and meeting with three more brethren, I inquired for father Roch in the ſame language: and one of them anſwered me only, *Padre Fratre Roch? ſuoras.* This was all I could get out of him. On this occaſion as on many others I felt how diſagreeable it was for a traveller to be unacquainted with the language of the country he is in. The want of comprehending exactly what others ſay; and the being obliged to be ſilent, when one has ſomething of conſequence to ſay, for fear of not being rightly underſtood, is a ſituation worſe than that of a deaf man, who has at leaſt the ſatisfaction of ſpeaking, and making himſelf underſtood.

On returning a ſecond time to the convent, I had the good fortune to meet with the ſuperior, who anſwered me in good Latin, which he ſpoke, though with ſome little heſitation. He brought me into his cell, where we converſed together for a full half hour, after which father Roch came and joined us. In the courſe of this converſation, he informed me of ſome remedies, the ſucceſs of which he had ſeen in repeated experiments. I give the receipts of ſome of them here, that any perſon who thinks proper may have an opportunity of trying them.

Tooth-ach.

Extract from the fuller's thiſtle, a worm which is always found in it when it is ripe. Rub this worm between the thumb and fore-finger, preſſing it gently till it dies through weakneſs. One or other of theſe two fingers applied to the tooth will have, at leaſt for a whole year after, the property of removing the pain.

The Farcy in Horſes.

At the end of autumn collect the bearded protuberances, or kinds of cheſnuts belonging to the eglantine: bruiſe the worm you will find in them, and make the horſe ſwallow it in a glaſs of wine, or any other liquid; then cover him up warm.

A foundered Horſe.

Let him take one or two ſpoonfulls of common ſalt, in half a pint of common water.

Malignant Fevers.

Under each ſole of the patient's feet, apply a tench quite alive, without ſlitting them, or doing them any injury. Bind them on with linen rollers, take them off at the end of twelve hours, taking care if poſſible, not to inſpire the ſmell that comes from them, then bury them quickly, or throw them down the houſe of office; and the patient will ſoon recover.

Quinſey.

Take as many earth worms alive as will make up the ſize of an egg; put them between two pieces of thin muſlin, and apply them round the patient's bare throat. Renew the application every three hours for two days ſucceſſively.

Bleeding of the noſe.

Put into the patient's two noſtrils, or behind both his ears, a ſmall quantity of hair taken from the private parts of the ſex different from the patient; and the blood will ſtop almoſt inſtantaneouſly.

An infallible plaiſter for bringing out the ſmall-pox when it has been driven in.

Take ſome rye meal; mix it up with ſome rain water, ſome verjuice, a new laid egg, and half an ounce of orpiment finely powdered. Beat the whole well together, and ſpread it upon blotting paper. Sprinkle it with cloves in powder, and apply this poultice to the ſoles of the feet; it muſt be left there for the ſpace of four and twenty hours, then taken off, and thrown quickly into the fire.

Fluor

Fluor Albus.

Bruiſe the leaves of the plant called mouſe-ear, and ſqueeze out the juice to the quantity of two ounces, which the patient muſt take faſting, in a cup of broth, or a glaſs of white wine. The doſe muſt be repeated for ſome days ſucceſſively, after ſome purging phyſic has been given to the patient; who will feed only upon meats of eaſy digeſtion, and will abſtain from all kind of exceſs. The father aſſured me that this medicine had cured women in five or ſix days, who had been afflicted with this diſorder for eight or ten years.

Immoderate Menſtrual Flux.

Torrefy, upon a new earthen plate, or upon the fire-ſhovel well cleaned, as much hair, taken from the private parts of a healthy middle aged man, as one can hold between the finger and thumb. Reduce it to powder; and let the patient take it faſting in a glaſs of good red wine. For a ſuppreſſion of the menſes it muſt be taken in white wine. The remedy may be repeated a ſecond time.

Swelled Glands and other ſcrophulous Tumors.

Apply one or two dead plantain leaves to them. Renew the application, with freſh plantain, twice a day. At the ſame time, let the patient take, every morning faſting, a warm infuſion of walnut-tree leaves, in the ſame manner as tea.

Colic, and Stitch in the Side.

Let the root of the ſun-flower be applied under the armpit of the ſame ſide where the pain is. As ſoon as it is grown warm there the colic goes off. This application was tried with ſucceſs in an obſtinate ſtitch of the ſide.

Exoſtoſis.

Flatten a ball, which has killed an animal, and apply it immediately upon the part affected.

Palſy.

Palſy.

Boil ſome radiſhes in water, with a ſmall quantity of gin, and drink it for common drink. One may alſo put radiſhes into the ſoup inſtead of common ſoup herbs.

Ulcers.

Chew the dried treadles of ſheep, and apply them by way of of poultice to the ſore. Let the application be renewed morning and evening.

Cancer and Ulcers.

Put a large live toad into a new earthen pot, and over it put two ounces of rolls of ſulphur in powder. Lute the pot well, and calcine the whole. Apply the aſhes to the cancer.

Corns and Warts.

After having ſcratched them and taken off the hard part, rub them well with the muſhrooms which grow naturally upon a dunghill.

Pains after Child-birth.

Boil two new laid eggs, and put into each of them a piece of ſugar in powder, as big as a filbert; mix it well with the yolks, and let the lying-in woman take it, drinking over it a glaſs of good wine daſhed with a little water.

To promote the Diſcharge of the Lochia.

Put two drachms of flower of ſulphur into two glaſſes of boiling water, let this boil for a few minutes, then ſtrain it through a linen rag, and let the liquor be taken.

An Amulet againſt the Falling Sickneſs.

Put into a crucible, upon a ſlow fire, one ounce of Spaniſh mercury, or mercury ſeparated from cinnaber. When the mercury acquires a little heat, and begins to ſimmer, throw in one drachm of ſilver beaten very thin, and ſtir the whole well with a rod of iron, a little heated. Afterwards remove the crucible quickly

quickly from the fire, and pour out the contents, leaving them to grow cold, Put this amalgama into a ſmall leathern bag, cloſely ſowed up. Faſten this bag round the neck with a ſtring, ſo as it ſhould fall upon the pit of the ſtomach, where it muſt be conſtantly worn. Before this amulet is applied, the patient muſt be blooded in the cephalic vein, at the new moon. The bleeding muſt be repeated, at the new moon, the two following months.

Bronchocele.

Apply ſome common ſalt, well dried and a little warm, to the tumor. When the ſalt grows damp, let it be taken off, well dried, and then applied again; this proceſs is to be repeated till the diſorder is cured.

Specks on the Eyes.

Take of dragon's blood, of ſuccotrine aloes, and of myrrh, equal quantities, and let them all be very finely powdered. Mix a ſufficient quantity of this powder with the yolk of a new laid egg, ſo as to form a plaiſter, which is to be applied to the temple on the ſame ſide as the eye affected. When the plaiſter falls off of itſelf, put another on, and continue in this manner till the cure is completed.

Pain in the Teeth, and how to make them fall out without Pain.

Put into the hollow of the tooth three drops of ſpirit of ſal ammoniac, and a ſmall bit of cotton over it.

Corns in the feet.

Take off the indurated part, without making them bleed, then apply ſeveral times, the red ſediment found at the bottom of a chamber pot, when the urine has been left any time in it. Then cover them with a piece of thin leather, repeating this till the corns are removed.

Fluxion

Fluxion in the Breaſt.

Set a pint of good cow's milk upon the fire; when it boils ſkim it two or three times, then throw in a large glaſs of good Spaniſh wine, and after it has boiled up twice, take it from the fire. When the milk is turned, ſtrain off the whey through a linen rag, and let the patient drink a ſmall glaſs full of it warm every quarter of an hour.

To bring about the Delivery of a dead Child.

Take ſome of the ſeed of the greater burdock, reduce it to powder, and let the patient take one drachm of it in a glaſs of wine.

Convulſions in Children cauſed by Teething.

Cut ſome of the root of wild valerian into ſmall pieces. String them like the beads of a necklace, ſo as to make a necklace of them for the child; who is to wear it 'till the teeth have pierced the gum. The application may be renewed every fortnight.

Dropſy.

Let the patient take, faſting, as much of the miſletoe of eglantine in powder as will cover a farthing, after it is infuſed the whole night in a glaſs of white wine, which is alſo to be taken. Half of this doſe only is given to women and children. This medicine was communicated to me by a Lieutenant of our frigate named Le Roy. He told me his father had tried it ſeveral times, and with ſucceſs.

Hyſteric Vapours.

Rub the inſide of a ſaucer with garlic laid on very thick. Then apply the ſide rubbed with garlic to the navel. Hold it on 'till it ſticks, and let it not be removed till it falls off of itſelf.

Fiſtulæ

Fiſtulæ of all Kinds.

Take the leaves of St. John's wort, of the leſſer wormwood, and of the round birthwort, of each one handful: ſuccotrine aloes, and myrrh in powder, of each one ounce. Let the whole be infuſed in two quarts of good white-wine, in a pot well glazed, and ſtrongly luted upon a gentle heat, for three quarters of an hour: let it afterwards boil for a quarter of an hour: ſtrain off the liquor when cold, and add to it one pint of good ſpirit of wine. Keep the whole in a bottle well corked.

This liquor is to be injected into the fiſtula five or ſix times a day, applying a tent or compreſs dipped in the ſame to the wound. This remedy has been tried ſeveral times ſucceſsfully, by M. Duvernay, a ſurgeon of Chambery.

For Diſorders of the Eyes, even the Gutta Serena, an Opthalmic wonderful in its Effects.

Take thirty-one live cray-fiſh, caught preciſely when the ſun and moon are in Cancer, and not at any other time. Take alſo of the roots, ſtalks, leaves and flowers of the celandine, gathered before ſun-riſe, as much as will equal the weight of the cray-fiſh. When all this has been well pounded together in a wooden or ſtone mortar, add of fennel-ſeed one ounce, of bean flower and camphor each half an ounce; cloves, hepatic aloes, prepared tutty, all in powder, of each two drachms. Mix the whole well in a mortar, and divide it afterwards into three parts. Put one of the parts into an alembic, and diſtil in B. M. till it is dry: take out the reſiduum, preſerve it, and put a ſecond part of the compoſition into the alembic, together with the water drawn off from the firſt diſtillation. Diſtil this again till it is dry. Take out the reſiduum a ſecond time, keep it, and put in, in its ſtead, the third part of the compoſition, with all the water diſtilled. Let the diſtillation be repeated a third time. Afterwards calcine the three reſiduums in a cloſe veſſel, extract the ſalt by diſſolution, filtration, and evaporation, *ſecundum artem.* Let the ſalt obtained be added to the diſtilled water, and after the whole has digeſted on a ſlow fire, keep the liquor in a bottle well corked.

Method of uſing the foregoing Application.

Let the patient be purged at leaſt twice with a mild cephalic medicine, leaving the interval of a day between the two doſes; and if he is plethoric he muſt be blooded once at the decline of the moon. Two or three drops of this collyrium are afterwards to be introduced into the eye morning and evening, with the black feather of a fowl s wing, and let a ſlight compreſs dipped in the liquor be applied over the eye.

During the uſe of this remedy, which muſt be continued about forty days for a gutta ſerena, one muſt be careful to keep the body open. For this purpoſe, if it ſhould be neceſſary, glyſters may be uſed, compoſed of river water only warmed. All melancholy things, and employments of too ſerious a nature, are alſo to be avoided; and the patient muſt likewiſe abſtain from ſpices, ſalt meats, ſellery, ſtrong liquors, and in general from all exceſſes in eating, drinking, ſitting up, &c. For other diſorders in the eyes, the collyrium is to be continued till the cure is completed.

An excellent and almoſt univerſal Balſam.

Put into a glazed earthen pan, which will bear the fire, and which holds about five or ſix quarts of water, three pints of fine olive oil, half a pound of freſh yellow wax, cut into ſmall pieces, half a pint of roſe water, three pints of good red wine, and two ounces of red ſaunders in powder. Let the whole boil together for half an hour, ſtirring the mixture all the while with a wooden ſpatula. When this is done, throw in a pound of fine Venice turpentine, not of the common ſort, with four ounces of good honey, and two drachms of camphor in powder. The finer ſort of Venice turpentine is not ſharp to the tongue, and has no diſagreeable ſmell; it is white and not yellow. M x the whole together by ſtirring it well with the ſpatula

ſpatula for one or two minutes: take the pan off the fire, ſtrain the balſam through a linen rag, and let it be kept in earthen pots.

Uſe of the foregoing Balſam.

For wounds, ulcers, mortifications, contuſions, burns, rheumatic and other pains, the part affected is firſt either waſhed or fomented with a little warm red wine; afterwards the part is very freely anointed with the Balſam, and a piece of blotting-paper ſteeped in it is applied. This dreſſing is repeated morning and evening. If the wound penetrates any of the cavities of the body, the balſam is to be thrown up with a ſyringe, and the patient muſt take a drachm and a half, or two drachms of it, in each baſon of broth, or in ſome decoction of vulnerary herbs. The ſame doze of this medicine may alſo be taken in the pleuriſy, the cholic, and other internal pains, taking care at the ſame time to apply it warm externally, rubbing it in on the part where the pain is ſeated. I have tried this Balſam and always with ſucceſs.

For an inveterate Head-ach, cauſed by a Fluxion of Humours, and for an Hydrocephalus.

Pound, in a wooden or ſtone mortar, ten or twelve tops of vervain, with ſome rye-meal, and five, or ſix, or more whites of eggs: the vervain may be omitted. Make a cataplaſm of this, which muſt be applied to the nape of the neck, and over the ſhoulders, ſo as to cover almoſt the whole ſcapula. Let a fine napkin four times double be laid over it, and let it be left on for ſix or eight hours. If the patient is not then cured, a ſecond poultice of the ſame kind is to be applied, which is to be left on as long as the other, or thereabouts. It very ſeldom happens that a third poultice is neceſſary. The patient muſt afterwards be purged. This poultice is equally beneficial in rheumatiſms.

A moist Asthma, Colds, and Disorders of the Breast.

In an earthen or very clean copper pan, let one pound of ripe juniper berries, well bruised, be boiled for half an hour, with a pound of fresh unsalted butter, which has not been washed. Then let the butter be strained off, with a very strong expression of the juniper berries. To the quantity strained off, add an equal weight of the best honey, and let it be put upon an exceeding slow fire 'till it has acquired the consistence of a syrup, which is to be preserved in earthen pots. The bigness of a small nut, or the value of a tea-spoonful, is to be taken in the morning fasting, suffering it to melt in the mouth like a lozenge. The same doze is to be repeated at night going to bed. When the disorder is dangerous, a third dose may be taken three or four hours after dinner.

In common disorders of the breast the juniper berries may be left out.

All these remedies have not been communicated to me by the Franciscan father whom I have before mentioned; but having seen the good effects of almost all those I have given before, I have been very glad of this opportunity of making them known for the good of the public.

The day after our party of pleasure into the country, which I have spoken of, four Indians or Natives of the country came to present themselves to the Governor, while we were with him examining Captain Mandillo's compass. As soon as the Governor saw them coming into his court-yard, he had all the doors of his rooms shut up. Upon our asking him the reason of this, he told us, that the room would be infected for eight days, if they were suffered to come into it; and that the smell which exhaled from them fixed itself even on the walls. This smell proceeds from their anointing their bodies with a certain kind of oil and grease to preserve themselves from insects.

These Indians finding the door shut, came up to the window where we were, and one of them pulled out of a bag made of a tiger's skin, a paper written and folded up, which he presented.

An Indian of Montevideo.

The Governor received and read it, being written in the Spaniſh tongue. It was a certificate, in which ſeveral Spaniſh Governors declared ſucceſſively, that the bearer of it was of the race of their Caciques or Princes, and that he himſelf was the chief of a village. The Governor returned the certificate, and the Indians aſked him by ſigns for a ſheet of paper inſtead of the one which before held the certificate, and which was now worn out in the the folds by much uſe: the paper was immediately given to him. It is moſt probable that theſe Indians were unacquainted with the Spaniſh language, as they did not attempt to ſpeak a ſingle word of it. A Spaniſh officer told us, that they had ſpoken the Paraguay language, mixed with that of the Indians, in the neighbouring parts. They had no other dreſs than a kind of cloke made of ſeveral deer ſkins with the hair on, ſewed together, ſo as to form a long ſquare pretty much like a napkin. It is faſtened about the ſhoulders with two ſtraps, and appears as in the plate. The ſide next the ſkin was white, and painted red and blue grey in ſquares, rhombuſſes, and triangles, the diſpoſal of which forms various compartments, according to the fancy, I ſuppoſe, of the perſon who is to wear it, or of the painter. Theſe Indians often come into the town in companies of five, ſix, eight or ten, bringing their wives along with them. Their dwelling-places are not more than ſix or ſeven leagues diſtant from Montevideo, where they come to drink wine or brandy. As they have no coin among them, they give their little ſacks of tiger ſkins, their clokes, ſometimes the ſkins of wild beaſts they have killed, but more commonly thoſe they have ſewed together to cloath themſelves with. They give them almoſt for nothing; for they exchange one of theſe kinds of clokes, compoſed of eight deer ſkins, for a real, which is about ſix-pence Engliſh. A ſack of tiger's ſkin, fourteen or fifteen inches long and twelve inches wide, coſts no more than half a real. When any body has a mind for the clokes of theſe Indians, it is ſufficient to take hold of it with one hand, and to offer a real or half a real with the other. The Indian immediately unties the ſtrap, takes the piece

of

of money, ſurrenders his cloke or little ſack, which ever you want and goes immediately quite naked to the firſt ſhop he can find to drink wine or brandy.

Their wives do the ſame. They have commonly no other clothing than the men, but now and then one ſees ſome of them who have a piece of the ſame ſkin their dreſs is made of faſtened round the waiſt with a ſtrap.

It is prohibited to ſell them ſuch a quantity of wine or brandy as will make them drunk, for fear that drunkenneſs ſhould lead them into ſome irregularities. M. de Bougainville being about to give a real to each of the four who came to the Governor's, the Governor deſired him on this account to give them only half a one. Being one day at the vicar's, we were told that a company of eight or nine of them, men and women, were coming up to us. The clerk of our frigate immediately going to the door with a bit of bread he was eating, one of the Indians paſſing by took this bit of bread from him, ſtopped a moment, then eat it laughing, and afterwards went on to join the reſt of his companions without ſaying a word. They were all of them bareheaded, their feet were naked, and they had no other clothing beſides the cloke already ſpoken of. Some of them wore it upon their right ſhoulder, leaving the left arm and ſhoulder bare; others wore it on the oppoſite ſide. They wear the hair on the outſide when it rains, and on the inſide when it is fine weather.

Such of theſe Indians as I have ſeen, were perfectly ſtrait and well-made, their arms and legs were well ſhaped, the cheſt well expanded, and all the muſcles of their body ſtrongly marked out. The women were much leſs than the men, who were all of a fine ſize. The women had, as well as the men, a lively look, a round but not a full face, pretty large eyes full of fire, a high forehead, a large mouth, and a wide noſe, flattened a little at the tip; their lips are of a moderate ſize and their teeth white; their hair long, black, and harſh, falling careleſsly about their necks, and ſometimes even over their foreheads. As they greaſe their

their hair as well as their body with different ointments, it is gloſſy; but always in diſorder. Theſe ointments however have nothing in them more diſagreeable or dirty than the pomatums of this country.

It is ſaid that when they are firſt born they have not that red, copper, bronze colour, which is generally ſpread all over their ſkin. It is true indeed, that the climate, the action of the air to which their ſkin, not covered with clothing, is continually expoſed, the ointments and paint with which they ſmear their whole body, may contribute, at leaſt in a great meaſure, to give them this colour. But when we conſider that the Negroes themſelves alſo do not come into the world with their ſkin black, which however is their proper colour, we may well imagine that the red copper colour of the Indians in South America is alſo natural to them.

The women are employed in the culture of manioc, and preparing it to make the caſſavi, and their common drink; they are alſo employed in houſehold affairs, whch conſiſt only in ſowing together deer and other beaſts ſkins, which both men and women uſe for their clothing, and in preparing victuals for themſelves and the men, who ſpend all their time in hunting, fiſhing, and riding out on horſeback; and indeed they are moſt excellent horſemen. The old men preſide in each hamlet of huts, and ſtay at home with the young lads and girls, who have not yet acquired ſtrength enough for any laborious work. Their form of government conſiſts entirely in reſpecting their elders.

They are extremely dexterous at handling the ſling, and at the management of the lance and the bow: they ſeldom miſs their aim with the ſling even on horſeback and at full ſpeed. A fierce bull, a tiger, or any other animal, or even a man himſelf, though ever ſo watchful, can hardly eſcape them. As it is neceſſary that the halter, which is the name they give it, ſhould confine the animal they have a mind to ſeize, they puſh their horſe at him ſtrongly, ſo as throw the halter in ſuch a manner that the animal finds himſelf dragged away with ſo much rapidity that

that he has not time either to disentangle or defend himself. In their private quarrels with one another, they also use these slings, and a half lance. The only method of avoiding this sling, if in an open plain, is to lie down all along upon the ground, as soon as they take the instrument in hand, and to keep close to the earth. Another method of avoiding it is, by sticking one s self close to a tree, or to the wall.

These halters or slings are cut out of bulls hides. They twist this strap, and make it flexible by greasing and stretching it out, till it is reduced to half a finger's breadth. Nevertheless it is so strong that a bull cannot break it, and it resists more than a hempen cord would do, which would also be less flexible, and therefore less fit for this purpose.

One can hardly get the skins of tigers and other wild beasts any otherwise than from the Indians. Nevertheless they are not dear, though rather scarce at Montèvideo. One of the finest of them may be bought for two or three piasters. I bought a very beautiful tiger's skin of a middling size, sowed up in form of a bag, for a piece of eight. The Indians kill but few tigers, though they eat them; because they make use of these skins only for the little bags I have mentioned. In these bags they carry the cassavi root, which serves for their nourishment, and the heads of their arrows, which they do not fasten to the reeds, till they are going to make use of them. This head of the arrow has the figure and size of a laurel-leaf, when much lengthened out at the two extremities. They fix it into the reed by either end indifferently, because it is pointed and sharp at both extremities. These arrows are the more fatal, as the head of them, not being firmly fixed, remains in the wound, upon attempting to draw out the reed.

When they want to catch an animal in the sling, they ride after him at full gallop, holding the horses bridle in one hand, and in the other the sling, which they throw at the neck, legs, or horns of the animal. When it is a furious or wild beast three or four of them together ride after him, each laying hold of a limb

limb in the ſling, then ſeparating, one to the right and another to the left, the ſling is tightened by this means, and another of them comes up without danger, either to kill the animal with his half lance, or to tie him and bring him away.

The Indians have other methods of hunting, which are deſcribed in the accounts of ſeveral authors, eſpecially in M. Muratori's work on the Paraguay.

I ſhall take this opportunity of acquainting the public, that M. Muratori's book is entirely written from the memoirs furniſhed him by the Jeſuits or ſome of their people, who were certainly ſo much concerned as not to be willing to inform the public of all their tranſactions. Some Spaniſh officers of credit, ſent from the court of Madrid to Paraguay, in the time of the diviſion of the reſpective poſſeſſions of the courts of Spain and Portugal, have aſſured me, that all the pamphlets they had ſeen in that country relative to the conduct of the Jeſuits, whether reſpecting the Indians or the intereſts of theſe two kingdoms, were always written with a great deal of caution in regard to the Jeſuits. He alſo told me, that one of theſe fathers, among the chief in the country, had made the following anſwer in his preſence, to one of the Spaniſh general officers, who was expreſſing his aſtoniſhment at the obſtacles which the Jeſuits oppoſed to the diſpoſitions concerted and fixed upon between the two courts. "I have much more reaſon to be ſurprized, that theſe two kings ſhould make diſpoſitions for dividing a country which does not belong to them. We Jeſuits alone have conquered it; we alone have the right to diſpoſe of it, to keep and defend it, from all, and againſt all." With ſuch principles as theſe, one may eaſily imagine what the conduct of the Jeſuits would be. It is certain that the Indians of Paraguay are ſubjects only to this ſociety, either at home in their families, or when they go out in arms. When the Spaniards lately beſieged and took from the Portugueſe the colony of St. Sacrament, which is about thirty leagues diſtant from Montevideo, they were aſſiſted by about a thouſand Indians, at whoſe head was a Jeſuit, who commanded them in

 chief,

chief, and without whofe order, thefe Indians would not have advanced one ftep, nor have fired a fingle mufket. The Governor of Montevideo, who commanded the Spaniards, and feveral other officers prefent at this attack, told me, they were obliged to fettle the plan of operations with the Jefuit, who afterwards gave out his orders in his own name to the Indians, who were encamped feparately from the Spaniards.

Dragoons are almoft the only troops of that country. Their horfes are equipped in the fame manner as at Paraguay. All the men wear the *Ponchos*, which they find more convenient than the cloke both for the horfe and his rider.

The Ponchos, as I have faid before, is a piece of ftuff formed like the coverlet of a bed, two or three ells long, and two ells wide. One muft pafs one's head through a flit in the middle, to put it on. It hangs down on both fides, and behind as well as before. It is wore on horfeback and on foot. The poor people and the negroes never take it off till they go to bed. It does not hinder them from working, becaufe it may be thrown back at the fides over the fhoulders; by which means the arms and the fore-part of the body are at liberty.

This kind of garment is fafhionable on horfeback, even for both fexes, and among perfons of all ranks. It is eafy, however, to diftinguifh ranks and fexes, notwithftanding the fimplicity of the Ponchos. Riding on horfeback is fo common among the women, that they are as ready and alert at this exercife as the men. The differences by which the rank and fex may be diftinguifhed with regard to the Ponchos, confift in the finenefs, lightnefs and richnefs of the ftuff.

The horfes are not fhod in this country. The faddle and furniture are alfo different from thofe ufed in Europe. They firft put upon the horfe s bare back a piece of coarfe foft ftuff, of a loofe texture, which they call *Schuaderos*; over that a girth, then a piece of ftrong leather of the fize of the faddle, which hanging over the horfe's crupper ferves for houfing. This is called *Carneros*. Over this leather is placed the faddle, made like that we

5 ufe

ufe for pack-horfes, and over the faddle they throw one or more fheep fkins, fewed together. This they call the *Peilhon*. Above all this they put a fecond girth, or furcingle, to faften the whole. The ftirrups are fmall and narrow, for they only put the end of the fhoe into them; and thofe who go bare-footed, reft only the point of the great toe. The bit of the bridle is iron, all of one piece, and without ftuds. The reins are compofed of feveral fmall ftraps, interwoven with each other, as the ftrings of a bell or clock; and are at leaft fix feet and a half, or feven feet in length, as they ferve at the fame time for a whip. A femicircular bar of iron, attached to the fame piece which receives the horfe's lower jaw, produces the fame effect as the curb. That part of the Carneros which projects beyond the faddle, and falls upon the crupper, is figured.

On the 9th inftant, the Governor, the major of the troops, and their ladies, came on board the Eagle frigate about noon, where we gave them as good a dinner as we could. The air of the fea, or the motion of the fhip, though fcarce perceptible, were rather troublefome to the major's lady, and made her fo fick, that fhe could neither eat nor drink any thing, except a couple of oranges, and was obliged to quit the cabbin where we were at dinner, to go and breathe the frefh air upon deck. This circumftance rather difturbed our entertainment, and obliged the company to return to the city very early.

As we were attending them to fhore in our cutter, we perceived an exceeding fœtid fmell, much refembling the putrid exhalation from the carcafe of an animal that has been dead a great while. We thought at firft that it had proceeded from the dead body of fome bull, killed and left upon the fhore till it was putrified, from whence the wind might bring it to us. The Governor undeceived us; affuring us, that it was the exhalation of the urine of an animal named *Zorillos*, who was either angry, or purfued by fome other animal.

The Zorillos is of the fize of a weafel, not quite fo long, with redd fh hair, lighter under the belly which is almoft grey. Two

 white

white lines extend the whole length of the back, forming, from the neck to the tail, almoſt an oval. The tail is very buſhy, and the animal always keeps it raiſed up as the ſquirrel does. When he finds himſelf purſued, or is provoked by any thing, he immediately expels his urine, which infects the air, to the extent of more than a mile and a half, with an almoſt inſupportable ſmell of carrion. We perceived this ſmell two or three times while we were on board our frigate, though we were more than four miles and a half diſtant from land: it is true indeed, that the wind blew from the land. M. Duclos, our captain, had already told us of this, but we had not taken his word for it. The fact was confirmed to us by the vicar of Montevideo, who made a preſent to M. Duclos of a fur lining made with the ſkins of this animal ſewed together. Theſe ſkins have no bad ſmell. The Zorillos is perhaps the ſame as the *Stinkbingſem*, or *enfant du diable* (devil's child) of Canada, the urine of which produces nearly the ſame effects. The *Chinche* of the ſouthern parts of America alſo reſembles much the Zorillos.

Another animal very common in theſe parts, and about Buenos Ayres, is the Tatu-apara, which we call *Tatou*, the Spaniards *Armadillo*, and the Portugueſe *Encubertado*. As this animal is very well known, I ſhall not give any deſcription of it. Ximenez ſays, that the ſcales of the Armadillo, reduced to powder and taken to the quantity of one drachm in a decoction of ſage, brings on a perſpiration ſo ſalutary, that it cures the venereal diſeaſe; and that it throws out ſplinters from all parts of the body: and according to Monades, liv. xv. pag. 552, the ſmall bones of this animal's tail cure a deafneſs.

Notwithſtanding the riſk there was in ſelling of any merchandize at Montevideo, and the difficulties our people met with in diſembarking them, to prevent their being ſeized; yet ſeveral of our officers and many of the crew, who had got together ſome few things, in hopes of ſelling them at the French iſland, and at the Eaſt Indies, where they thought they were going, got rid of them, and were ſelling them every day.

day. Our ſhip being the firſt veſſel which had put in at this place ſince the peace, our things ſold very well. The guards ſeized upon ſome packets brought on ſhore with too little caution. They even threw yeſterday into priſon two cockſwains, on whom theſe packets were found. M. de Bougainville being informed of this, exclaimed and ſtormed very much againſt theſe cockſwains, ſaying, they deſerved to be treated as they were. He went to the officer, repeated the ſame thing to him, and deſired only that his uniform, which the priſoners had on when they were taken up, might be delivered; that as for the men they might keep them, to intimidate the reſt, and that he himſelf would put them in irons if they were releaſed. By this behaviour, M. de Bougainville perſuaded them that he did not countenance this practice. The parcels and the men were both given up, and it was even deſired that the men might be excuſed. From this circumſtance, it was eaſily found out that our people had not taken proper meaſures. A ſerjeant having complained on this occaſion, that he had not received the value of one real, though he had aſſiſted in bringing many parcels on ſhore, and that he had, as he ſaid, wore out a pair of ſhoes in going about the town to ſhew ſuch of our people as had any thing to ſell, into the proper houſes; this declaration made us underſtand that we ſhould not meet with ſo many difficulties, if we did but diſtribute a few piaſters among the officers and guards. We began therefore by giving away a few pieces of eight, a few ſhirts, &c. and every body was very ready to let us do as we pleaſed, even the officer himſelf placed there on purpoſe to prevent any kind of trade. As we were ſuppoſed not to have any Spaniſh coin, and that French coin was not current in that country, M. de Bougainville aſked and obtained leave to ſell ſome butts of wine, brandy, oil, and ſome other ſuperfluous goods he had, in order to pay for the flour, and the freſh meat, the Governor had always taken care to provide for us every day at the port; as well as for the oxen, cows, horſes and other animals, he wanted to buy. It was now time to think of quitting

ting Montevideo, in order to go to the place of our deſtination. We therefore provided ourſelves with every thing we thought neceſſary for this purpoſe: the quantity of water, flour, and animals M. de Bougainville aſked for, excited the curioſity of the Governor with regard to the deſign of our voyage. All the crew, even the officers themſelves, were perfectly ignorant as to this point, and thought, as I ſaid before, that we were going to the Eaſt Indies. This report had been propagated, and M. de Bougainville confirmed it, by telling the Governor we were going to the Indies, without ſpecifying to which of them.

The trouble neceſſarily attending theſe preparations prevented us from day to day from trying to make any accurate obſervations with the inſtrument or compaſs of captain Mandillo, ſo that we did not think of fetching it from the Governor's till we were juſt preparing to ſet ſail. As it is of the greateſt conſequence to take advantage of the firſt favourable wind, eſpecially in the harbour of Rio de la Plata, where the anchorage is ſo dangerous; and as we ſaw that this compaſs could not be of much uſe to us with regard to the longitude, M. de Bougainville reſolved to leave it with the Governor. He wrote to him by the captain of a ſchooner, to deſire that he would keep this inſtrument, and at his return into Spain * ſend it to him in France. All the reſt of the time we ſtaid in the harbour, was employed in preparations for our voyage. Our frigates were carefully inſpected, and every poſſible precaution taken. Having brought on

* The Governor had told us, he expected to go back into Europe towards the end of the year; but we learned, on our arrival at Paris, that he would not quit his government ſo ſoon. M. de Grimaldi, the Spaniſh ambaſſador in France, aſked M. de Bougainville ſeveral queſtions, upon this Governor's behaviour to us. M. de Bougainville having given by his anſwers a proper teſtimony of the probity of Don Joſeph Joachim de Viana, and of his ſtrict obſervance of the duties of his ſtation; the ambaſſador confeſſed, that the Jeſuits and their friends had ſent over to Madrid memorials againſt him, to injure him in the King's opinion, that he might be recalled, and that they might have ſome Governor devoted to them in his ſtead. M. de Grimaldi juſtified Don de Viana to his court; which probably prevented the King from recalling him as ſoon as he wiſhed. The public news-papers have informed us of this gentleman's being continued in the ſame government.

on board twelve cows, or heifers, ſix mares, two colts, and two Hungarian horſes, twelve goats or kids, eleven ſows, and one boar, fourteen or fifteen ſheep, two rams, and a great quantity of fowls and ducks, we ſet ſail on Tueſday the 16th of January 1764.

At three in the morning, the wind, which for ten days paſt had been conſtantly South Eaſt, ſhifted to the North. The gale being moderate, we took this opportunity to unmoor. We heaved apeek to our beſt bower anchor, ſhipped the longboat and other boats, one of which had landed the Sieur *Sirandré* Lieutenant, with a letter of thanks to the Governor, from M. de Bougainville and the reſt. At nine o'clock we ſet ſail with the Sphinx and the Spaniſh frigate the St. Barbe, which we ſoon forereached, though ſhe was at leaſt two full leagues and a half ahead of us. We ſteered half a league S. E. $\frac{1}{4}$ S. about as far S E. then directed our courſe S. E. $\frac{1}{4}$ E. in order to double a point, which conſiſts of a chain of rocks on the South Weſt of the fortreſs, extending near a league out in the river. When we ſet ſail it blew a pretty freſh gale at North Weſt. The wind abated gradually and a calm ſucceeded; inſomuch that, at half an hour paſt three in the afternoon, the ſhip making no way, we anchored in ſix fathom and a half water, muddy bottom. The Sphinx, together with the St. Barbe, anchored at the diſtance of a long muſket ſhot from our ſtern. During the calm we caught three very beautiful butterflies, particularly one delineated in the plate *.

Our anchorage in the road of Montevideo was not abſolutely bad; but I am of opinion it would have been better higher up in the bay. During the whole time that we remained there, we were conſtantly on the alarm, as well on account of the *Pampéros*, which almoſt always riſes on a ſudden, as of the South Eaſt and South Weſt winds, which blow full into the mouth of

* I gave it the name of the parrot, becauſe the various colours of its wings exactly reſemble thoſe of the moſt beautiful parrot of Brazil. Its body is of the fineſt green, ſtreaked with red.

the

the bay and occaſion ſo high a ſea as to make it impoſſible to leave any of our boats along-ſide the veſſel. We were every night obliged to ſhip them.

The little boat being only once forgot, it got looſe and had like to have been loſt, as I have before related. When you are farther within the bay, you have nearly the ſame depth of water, the ſame bottom, and are ſheltered by the mountain on one ſide, and the town on the other.

We ſet ſail about eight in the evening, and continued our courſe the next day without any remarkable occurrence.

On the nineteenth, near two o'clock in the afternoon, an extraordinary kind of fiſh paſſed near the veſſel. We had ſeen a great many before; but they being at too great a diſtance from the ſhip, we had not been able to catch any of them. Perceiving that they aſſembled to-day round the ſide of the frigate, I got a ſailor to throw a bucket faſtened to the end of a rope into the ſea, and he had the good fortune to take one. Our mariners give them the name of *Galere*, or ſea-nettle. It is a kind of bladder, which may be ranked under the ſame genus with what the naturaliſts call *Holoturiæ*, which, without any reſemblance either of a plant or a fiſh, are nevertheleſs, really poſſeſſed of life, and tranſport themſelves like animals from place to place with a motion peculiar to themſelves, independently of the aſſiſtance of wind and waves on which you ſee theſe bladders carried like ſmall veſſels. Any one who did not obſerve this appearance of a bladder with a nice and judicious eye, would take it for a bubble of air floating on the ſurface and driven by the waves and winds. But the ſailor who had caught it having brought it to me, I had ſufficient time to examine it. I obſerved in it a periſtaltic motion, ſuch as anatomiſts aſcribe to the inteſtines and ſtomach. I was juſt on the point of taking it out of the bucket with my hand, when M. Duclos our captain caught hold of my arm, and bid me take care leſt I ſhould ſoon have reaſon to repent of ſo doing, by the acute pains I ſhould feel in every part of my hand, which ſhould happen to come

come in contact with the filaments, of a violet blue colour attached to this bladder. I therefore contented myself with viewing this sea-nettle and taking a sketch of it.

The captain's observation was verified the same day. A cabbin-boy having caught another of these fish, had the imprudence to take hold of it with his hand. The filaments, twisted themselves round it. The moment after, he began to cry out that he felt a smarting and very painful heat on all the back of the hand, and the wrist. He shook it immediately to get rid of the fish; but it was now too late. His cries hastened us to his assistance; he wept, and stamped, complaining that his hand seemed to be in a fire. It was bathed in oil; a compress dipt in the same liquid was applied to it, but the pain still continued more than two hours; when it went off gradually.

The sea-nettle is an oblong bladder, flattened underneath, rounded in its circumference, and blunted as it were at its extremities; from whence proceed those filaments, the touch of which occasions so much pain. One of these extremities is more rounded than the other; which is rather lengthened. The part which forms the base or resting point of this bladder is plaited about the edges. The whole is a membrane of a very delicate structure, transparent, and nearly of the same figure as those half globules, which rise on the surface of the water in summer showers, especially when they fall in large drops. It is always empty, but distended like a football. This membrane has fibres, some of which are circular, others longitudinal, by means of which the peristaltic motion is carried on.

At the longest of its extremities it contains a small quantity of the clearest water, which is prevented from communicating with the rest of the cavity by a membranous partition. The fibre which passes over the back, from the fore to the hind part, is raised, scolloped at the edges, plaited like a beautiful tuft, of a lively green, blue, and purple colour, extended in the form of a sail. It lowers, elevates or shifts, as it were to set itself for the wind. From the two extremities of the plait proceed some filaments of different lengths, and of the ſame colour with this kind of ſail: two of them are very ſhort, and as thick as a large quill; theſe afterwards branch out into ſeveral others of leſſer thickneſs, but much longer; and theſe again into others, ſtill longer and ſmaller, to the number of eight in all. They are about a foot long; but not all of equal length. Theſe ſtrings, interwoven with each other, reſemble a net whoſe maſhes are of different ſizes. They have a kind of articulation, formed by ſmall circular rings, in which one may likewiſe obſerve a contractile motion. Theſe filaments reſemble looſe taſſels, compoſed of ſtrings of a ſky colour tinctured with purple, and of a greeniſh caſt, nearly tranſparent, and of different lengths, the edges of which appear to be indented, and alternately intermixed with grey violet and flame colour.

The largeſt ſea-nettles I have ſeen were about ſeven inches long at their under part, and five in height. It would be very difficult to determine preciſely the colour of this extraordinary fiſh. The bladder is as clear and tranſparent as the fineſt chryſtal; but its edges, back and legs may be ſaid to contain the colours of the rainbow, or of the flame of ſulphur. We ſaw a great number in our paſſage, particularly in the ſtreight which forms the iſland of St. Catherine, at Brazil; and I believe they are common in theſe latitudes. If the bare touch of this animal cauſes ſo much pain, what can we think its effect muſt be in the bodies of fiſh or other animals who feed upon it? It has this ſurprizing quality, ſays Father Labat, that it taints and poiſons the fleſh of fiſhes without occaſioning their death.—This is pretty nearly the effect of the fruit of the manchineel-tree.

On the 20th we perceived that the currents ran to the S.S.W. which confirms the obſervation in the account of Admiral Anſon's voyage. The tides carried us 30 min. Southward; and there is reaſon to believe that their courſe is to the S.S.W. agreeable to the bearing of the coaſt.

On Sunday the 22d in the morning the wind, which the day before had blown very freſh till four in the afternoon, grew calm, with

The material originally positioned here is too large for reproduction in this reissue. A PDF can be downloaded from the web address given on page iv of this book, by clicking on 'Resources Available'.

with fair weather and a fine ſea. We ſaw ſome *Dadins*, ſea-fowls which are found common in almoſt all latitudes, and ſome large birds called *Quebranta-hueſſos*, as well as ſome *King-fiſhers*, which our mariners call likewiſe *Puans*. 'Tis ſaid that when theſe laſt appear, there ſeldom fails to be foul weather, and often ſtorms, either on that or the next day.

In fact, the S. S. W. wind we had, ſoon after blew with violence: the ſea ran high, the weather became foggy, and it rained a little at intervals.

In the evening of the 23d we ſaw ſeveral birds, and ſome very long and large and well formed beds of fiſhes fry, of a reddiſh caſt. Moſt of them extended beyond our ſight in length, and ſome were about a hundred feet in breadth.

On the 24th we ſaw eight or ten whales, a great number of birds, and a kind of ſea-weed, which our mariners called *Baudreu*.

On the 25th the wind blew very freſh from the N.W. till five in the evening. The rolling of the ſhip was ſo conſtant and violent, that we loſt a goat, two ſheep, and three cows. Several others fell ſick, as well as the horſes we took on board at Montevideo.

The weather grew dark and rainy. At ſix o'clock the wind rather abated, and ſhifted to the Weſt, then to the Weſt South Weſt, to the South, a gentle gale. The ſea likewiſe gradually ſubſided.

On the 26th we obſerved that the tides turned towards the North. We again met with ſeveral birds, and ſome ſea-graſs. The ſea ſtill continued much agitated till ſeven in the evening. We were obliged to kill one cow and a goat, taken ill with the rolling of the ſhip. The ſea ſoon after grew calm, and continued ſo almoſt the whole night. In the evening of the 27th we ſaw a quantity of birds, among which were ſeveral king-fiſhers. The wind blew with great violence, and the weather became dull and foggy. This laſted almoſt the whole night, and deſtroyed

 a very

a very fine ſtallion, whom we were obliged to throw overboard, as well as a goat and a ſheep.

On Saturday the 28th we ſaw a whale, two ſea-wolves, and two penguins; in the afternoon there appeared great numbers of oſpreys, and ſea-cobs, and we met with ſome ſea-graſs, with long leaves. On Sunday the 29th, about three in the afternoon, ſome pretty large fiſhes appeared near the ſurface of the water. Several of our mariners who have been uſed to fiſh at Newfoundland, aſſured us, that they were ſtock-fiſh.

On the 31ſt, at ſix in the morning, we deſcried land at the diſtance of about ſix leagues. We took this land for iſlands. We had then a ſtrong gale, which abating about eleven, we ſteered E. S. E. till noon, when we made the ſouthermoſt point of land, bearing S. E. five degrees E. about a league diſtant. The moſt eaſtern point bore E. $\frac{1}{4}$ S. E. diſtance two leagues, and all theſe lands appeared to be iſlands.

The figure of theſe iſlands, which form a triangle, as the Sebald iſlands are ſaid to do, and the idea we entertained of our being near theſe, induced us at firſt to believe that theſe three iſlands we ſaw were actually the Sebaldes.

On this account, according to our obſervations at noon, we found them placed in Belin's French chart thirty leagues too far to the weſt; and we were the more deceived in our obſervation, on account of its agreeing with Father Feuillée's and with a manuſcript chart which M. de Choiſeul gave to M. de Bougainville, before we left Paris. See theſe iſlands in the plate as they appeared to us at two leagues diſtance, the Cape lying Eaſt-South-Eaſt.

This chart of M. de Bougainville's extends the eaſtern verges of the Malouine iſlands to 57 degrees 15 minutes longitude; and Father Feuillée places the ſame extremity of theſe iſlands in 57=45: the latitude agrees alſo pretty exactly. M. Belin fixes it at 62 degrees. We ſhall be better able to determine which of the two is in the right when we land, as we intend to do.

In

In the afternoon we had a fresh gale at N. W. As we kept coasting along the shore we sounded at three o'clock, at 45 fathoms, a flinty bottom. At four we sounded at 40 fathoms, flinty bottom mixed with broken shells, we were then half a league distant from two flat islands, which at first view appeared to be covered with small copse, but these were only tall bulrushes with flat and large leaves called corn-flags, as we discovered afterwards on landing; the coast being quite full of these corn-flags, as it now appeared to us. Sounded again, and found a rocky bottom at twenty-four fathom.

In the afternoon of the 31st, we coasted along the shore, at the distance of about a league and sometimes only half a league, in order to observe it with greater advantage. We sounded from time to time at thirty-five fathom depth, grey sandy bottom.

The lands are of a moderate height from the sea, with eminences, some behind others; a proof that this is either the continent, or the largest of the islands. Almost all the shores were covered with bulrushes which looked like small trees. This appearance is occasioned by the corn-flags growing each of them about two feet and a half high, and afterwards shooting forth a tuft of green leaves nearly of the same height. This we had an opportunity of observing more particularly when we landed. We saw no wood, and at this distance the soil of the country appeared parched and dry; perhaps the heat of summer might have withered the grass.

At three o'clock we saw a small island two leagues wide of the coast. It nearly resembled in figure that on which the *Fort de la Conchée* near St. Malo is built. M. Bougainville gave it the name of the Tower of Bissy *. At five, we discovered a Cape, and a small island, resembling Cape Frehel, situated four leagues from St. Malo. This Cape seemed to terminate the land to the East.

On

* This is the entrance of the streight which divides the island into two parts, the eastern and the western. This streight runs from North to South.

On the firſt of February, we perceived another Cape and a ſmall iſland almoſt ſimilar to thoſe which reminded us of Cape Frehel; and after that, another ſmall one intirely covered with birds. At noon, the wind blowing ſtrong with ſqualls and rain, cauſed ſo violent a rolling of the ſhip, that our cattle ſuffered much from it. At laſt we determined to kill ſeveral ſick cows, fearing they ſhould die, and we ſhould be obliged to throw them over-board, as we had the fine bull we had brought with us from St. Catherine's iſland, as well as ſome goats and ſeveral ſheep.

At ſix in the evening the weather being then fine, with a gentle breeze, we determined to ſend out the fiſhing-boat which was manned for that purpoſe. Meſſrs. Donat and Le Roy the lieutenant, went on board with a ſufficient number of ſeamen, all well armed. They were ſent on ſhore to cut graſs for our cattle, who began to be in want of it. We were then about two leagues from the point which appeared woody. We were becalmed till about eight o'clock. The tide drove us towards the ſhore upon a ſhoal of rocks. In this embarraſſing ſituation, from which it was impoſſible to extricate ourſelves for want of wind, we ſounded with a view of caſting anchor, if the bottom ſhould be good. But the bottom proving rocky at between eighteen and twenty fathom, our perplexity increaſed, and with the more reaſon as the tide had already carried us towards the ſhoal, which lined a pretty large creek, and we were ſcarce half a quarter of a league from it. The Sphinx laboured under the ſame difficulty, and we were already contriving means to ſave our lives in caſe we ſhould be ſhipwrecked upon theſe rocks, which the mariners call *the Carpenters*; becauſe a ſhip which has the misfortune to run aground here, is ſoon daſhed to pieces. Fortunately, about eight o'clock, a very faint breeze blew from the ſhore; and our officers, equally attentive and able to avail themſelves of the ſmalleſt advantage, ordered the working of the ſhip ſo ſkilfully, that we got clear of the ſhore. The ſhip's crew were ſo fully ſenſible of the danger we were in, that in the moſt tempeſtuous weather, and even during the ſtorm we ſuffered near the Maldonnades, they never worked the

the ſhip with ſo much alacrity and diligence. It was a fine ſight to ſee every one at his poſt, holding in his hand the ropes he was to manage: all, in an attitude, in which was pictured anxiety and fear mixed with hope; all, obſerving the moſt profound ſilence, their eyes fixed upon the captain, and their ears attentive to catch the firſt word of command: the two captains and the lieutenants, and all the ſhip's company, employed in looking, ſome on the ſide of the ſhip towards the ſea, others towards the land, to obſerve if any one could perceive the ſmalleſt breeze riſing, and ruffling the ſurface of the water which was almoſt as ſmooth as glaſs. One turned his cheek, another held his hand, and a third wetting his, extended it towards the quarter from which they imagined the wind began to blow in order to perceive the leaſt motion. At length the long wiſhed-for breeze aroſe, but blew very faintly; fear gave place to joy and ſatisfaction, and to prevent our being again involved in the ſame difficulties, we ſteered away North Eaſt ÷ Eaſt, five degrees Eaſt.

About eleven our fiſhing-boat returned loaded with greens, and was taken on board. Meſſrs. Donat and Le Roy informed us, that they had ſeen at land, about the diſtance of a muſket-ſhot from the place they were in, an animal of a terrible appearance and aſtoniſhing ſize lying upon the graſs; his head and mane reſembling a lion's, and his whole body covered with hair, of a duſky red as long as a goat's. This animal perceiving them, raiſed himſelf upon his fore-feet, eyed them a moment, and then lay down again; having afterwards fired at a buſtard, which they killed, the enormous animal raiſed himſelf a ſecond time, eyed them as before without changing his ſituation, and then lay down again. According to their account, this animal ſeemed to be as large as two oxen, and twelve or fourteen feet in length. They had a mind to fire at him, but they were terrified, and durſt not fire for fear of wounding him ſlightly and hazarding their lives; or, according to their own account, they were unwilling to loſe time, as it was late, and they were deſirous of returning on board.

On the third about noon we discovered an opening of a bay *, the entrance of which appeared ſo fine, that we went into it full ſail, as into a well-known and commodious harbour. We anchored about three leagues within the bay, which appeared to extend at leaſt as many leagues beyond us. At the fartheſt extremity we diſcovered larger and ſmaller iſlands, where the fiſhing boat ſounded at four, five, ſix fathoms and more, in a muddy bottom. The Sieur Donat having been ſent thither immediately after we came to anchor, informed us, on his return, at ten in the evening, that it was every where at leaſt between eight and ten fathom, and between ſeven and eight to the Eaſt of the iſland, with a bottom of muddy ſand throughout, which ſecured us a retreat in caſe of bad weather at ſea.

This bay, the plan and figure of which is given in the plate, is capable of containing at leaſt a thouſand veſſels, and as many more to the weſt of the large and ſmall iſlands, which are ſheltered from all winds, and are, as our mariners ſay, more ſafe than in the harbour of Breſt.

As ſoon as we had dined, we ſent out the yawl and long-boat, and Meſſrs. de Bougainville, Nerville, Belcourt, l' Huillier, Donat, Sirandré and myſelf, landed on the ſouth ſide of the bay. As we were going to ſhore, a prodigious number of black and white birds, of the ſame ſpecies, croſſed in large flocks, no more than five or ſix feet above our heads. We killed ſome of them. Thoſe which fell into the ſea wounded, only dived when we attempted to take them up. Before we landed, we ſhot buſtards, geeſe and ducks, which did not fly away when we approached them; but walked near us, as if they had been tame.

While we were at a diſtance from the land, its appearance deceived us. We expected to find the face of the country dry and parched, but on ſetting foot upon it, we found it entirely covered with herbage, or a kind of hay, a foot, or a foot and half high,

* See the entrance of this bay in the plate. It is ſituated on the eaſtern ſide of the Malouine iſlands; and may be ſeen at the diſtance of three leagues.

high, reaching even to the tops of the hills, which we had much ado to climb, on account of this hay obſtructing our paſſage.

We aſcended in companies, while ſome took ſeparate routs for ſhooting, both upon the hills, and along the coaſt. We were much fatigued in climbing theſe hills, there being no road or path through this herbage, which is probably coeval with the ſoil.

We walked up to the knees in this hay; and the ſoil, which appears to be a dark brown, is formed into a mould by the annual decay of the hay, and riſes with a ſpring under your feet, owing to the roots which are intangled with it. Hence it is evident, that one cannot walk for any time over ſuch a path without being fatigued. But we were luckily provided with ſome ſmall bottles of brandy, and ſome ſea-biſcuit, which were of great ſervice, as we ſhould otherwiſe have ſuffered from the intenſe heat.

Here we met with ſome green hillocks, raiſed ſometimes more than three feet from the ſurface. I examined one of theſe with attention, and found that a reſinous gum oozed from it, which is white at firſt while it is ſoft, and of an amber colour when it hardens. I gathered ſome pieces of it, which I perceived had at leaſt as ſtrong an aromatic ſmell as frankincenſe: but could not determine at that time the exact reſemblance this gum bears to other gums or reſins which have hitherto been diſcovered. I brought away near the weight of half a drachm in grains or drops, ſome of the ſize of a round pea, others of that of a kidney bean. When I returned on board, I ſhewed them to M. de Bougainville and our two ſurgeons. I held ſome of it on the point of a knife in the flame of a candle; it burnt like the fineſt reſin, emitting an agreeable ſmell, and leaving behind it a blackiſh oil, which did not burn, but grew hard and brittle when it was cold. I attempted to diſſolve this oil in common water, but without effect: from which I concluded, that it would be very proper to make an excellent varniſh. Having mentioned it the next day to M. Frontgouſſe, ſurgeon of the Sphinx, he went on ſhore, and having collected ſome of this gum, conjectured from its ſmell and taſte that it was gum ammoniac. On comparing theſe, we found they agreed in taſte and ſmell, and were reduced to the ſame ſubſtance after burning. The ſmell is retained ſo ſtrongly on the fingers, that tho' I waſhed my hands more than once with ſea-water, I could not get rid of it all that day and the next. This reſinous gum only diſſolves partly in ſpirit of wine, which it tinges with an amber colour. The reſiduum becomes ſpongy and burns as it did before it was diſſolved: the third reſiduum does not diſſolve in common water. Aqua-fortis has no effect upon it.

Theſe hillocks are formed by a ſingle plant, from which proceed ſmall ſpongy ſtalks gradually dropping their leaves like the palm-tree. Theſe leaves are tripartite, as may be ſeen in the plate, where the plant is repreſented in its natural ſize. They are of the thickneſs of thoſe of purſlain, but of a fine green, ranged very cloſe to each other, diſpoſed in a circle, and forming a cavity in the center which is ſcarce diſcernible. This is a kind of very flat funnel, the inſide of which is lined with theſe leaves placed cloſe one above another, imbricate like thoſe of an artichoke. See the figure of it in the plate.

From the center and edges of theſe leaves, when they are bruiſed or only ſcratched, or when the reſinous juice is too plentiful in them, this reſinous gum, which congeals in the air, is produced. On cutting, ſcraping or even rubbing the ſurface, there iſſues a kind of cream which is white and viſcous, and ropes between the fingers ſticking faſt to them like glue. I have given it the name of the *varniſh plant*.

The inſide of theſe hillocks is formed into a vault, ſupported by ſtalks and branches, whoſe leaves, having no air, are brown and withered. Other plants ſometimes ſhoot forth on the inſide of this vault, penetrate through the hillock, and riſe above it. Theſe hillocks, when they are not broken, are ſufficiently firm, not only to ſupport a perſon ſitting upon them, but walking over them. The vault, however, is eaſily broken through, by ſtamping upon it with ſome force, and it is eaſy to tear off large pieces with one's hands. The root and the ſtalks when

5 broken,

broken, yield the ſame white reſin, which flows out like the white juice or milk of the plant called Tithymalus, which I ſhall ſpeak of hereafter.

In the evening our ſportſmen returned loaded with geeſe, buſtards, ducks, teals, and a black and white bird already mentioned. I had ſeparated from my companions, and wandered alone along the coaſt a league beyond the place where the boat had landed. I ſhot ſome ducks four or five feet from the ſhore. As I did not care to go into the water, I imprudently drew them to me with the end of my piece. The plenty of game engaged me to load again without recollecting that the barrel might have taken water. The powder was ſo wet that it would not take fire: and having no ſcrew, I reſolved to repair to the boat. I had ſcarce advanced twenty ſteps, before I perceived in the graſs a path eight or nine inches broad, very much worn, which ran parallel to the ſhore at ten or twelve feet diſtance from the ſea. I then imagined, that the iſland was inhabited, if not by men, at leaſt by quadrupeds which frequented that ſpot. But as I did not know what theſe animals were, they might poſſibly be of the ſavage kind; and I was apprehenſive leſt I ſhould meet with ſome of them in my way. Unattended as I was, without any other defence than a muſket now become uſeleſs, I was rather anxious for my ſafety. I fixed my bayonet to the end of my piece, and purſued my journey in this tract, being deſirous to know where it terminated. At the diſtance of about two hundred ſteps from the place where I entered it, it led into a thicket of thoſe kinds of corn-flags I have already mentioned. Not daring to venture into it, I ſtopt a few minutes as I paſſed near it: I obſerved it attentively, and liſtened to hear whether any thing ſtirred. I could perceive neither motion nor noiſe. I continued my march, and re-entered the path on the other ſide, till I diſcovered the boat, in which, our ſeamen perceiving that night was approaching, and that the ſeveral parties who had gone on ſhore to reconnoitre and hunt were not returned, came to meet, and take us on board. It was almoſt full of game, and the night

 obliged

obliged us to throw into the ſea a great part of what we had procured in the day.

On Saturday the 4th at ſix in the morning we got the yawl and fiſhing-boat ready to make ſome diſcoveries relative to the depth of the bay, which appeared to us to be a large river when we ſaw it from the heights the evening before.

Meſſ. de Bougainville, de Belcourt, de St. Simon, l'Huillier, and Alexander Guyot, embarked in the yawl, well armed, and furniſhed with proviſions for four or five days, and a tent to uſe on the ſhore. The ſailors too were armed with muſkets, cutlaſſes and bayonets. Their deſign was to proceed to the northern part, and to find out whether it was covered with wood. Meſſ. Donat and Arcouet, in the fiſhing-boat, were to make diſcoveries on the ſouthern part, an idea having been entertained that the bottom of the bay was divided into two branches which loſt themſelves in the vallies.

Meſſ. Alexander Guyot and Arcouet returned on board the ſame evening in the yawl, having left M. Donat with the fiſhing-boat to join M. de Bougainville and his aſſociates. M. Guyot brought ſome buſtards, three young ſea-wolves with hair of a browniſh grey, and five ſea-lioneſſes. They were about ſeven feet long, and three and a half in circumference, though their inteſtines were drawn. Theſe gentlemen had landed on a ſmall iſland, where they found a prodigious number of theſe animals, and killed eight or nine hundred of them with ſticks. No other weapon is neceſſary on theſe occaſions. A ſingle blow with a bludgeon, three feet or three feet and a half long, aimed full at the noſe of theſe animals, knocks them down, and kills them on the ſpot.

This is not altogether the caſe with the ſea-lions: their ſize is prodigious. Our gentlemen encountered two of them for a long time, with the ſame weapons, without being able to overcome them. They lodged three balls in the throat of one while he opened his mouth to defend himſelf, and three muſket ſhot in his body. The blood guſhed from the wounds like wine from

from a tap. However he crawled into the water and difappeared. A failor attacked the other, and engaged him for a long time, ftriking him on the head with a bludgeon, without being able to knock him down: the failor fell down very near his antagonift, but had the dexterity to recover himfelf at the inftant the lion was going to gorge him. Had he once feized him, the man would infallibly have been loft: the animal would have carried him into the water as they ufually do their prey, and there feafted upon him. In his retreat to the fea this animal feized a penguin and devoured him inftantaneoufly.

There are feveral kinds of fea wolves and lions; all which I have feen. The former, when at their full fize, are from ten to twenty feet in length, and upwards; and from eight to fifteen in circumference. Their fkin is covered with hair of a clear tan colour, or fallow like a hind's, and as fhort as that of a cow. The head is fhaped like a maftiff's, fuppofing the lips of the upper jaw were divided under the nofe like thofe of a lion of the foreft, and were not pendulous; and that the ears were cropped clofe to the head. I fhall defcribe them more fully afterwards.

The other fpecies, which is not fo large, has the fame appearance; the fnout is rather rounder and fhorter. Inftead of fore paws, it has two fins confifting of articulations, covered, as with a glove without fingers, with a very hard fkin or membrane of a dark grey colour. Thefe articulations are not diftinguifhable on the outfide, and can only be difcovered by diffecting the fin. The two hinder feet have vifible articulations like the fingers of the hand, five in number, and of unequal length. Thefe fingers from the firft to the third articulation are joined by the membrane: which afterwards divides itfelf, and runs along the fide of each finger, in the fame manner as the membrane in the feet of a diver or water-fowl, and extends much beyond each finger. Its feet are fituated almoft at the extremity of the body; where they form a kind of fplit tail when they lie down or are not in motion. Each toe is armed with a claw which

which is not ſharp, but rather projecting, and of a black colour. See the figure in the plate.

Both kinds are bearded like tigers, and have thick ſtrait hairs directly over their eyes by way of eyebrows. The female appeared to have a longer and more graceful neck than the male; and had dugs

In theſe animals the fat, which is white and flabby, is ſo redundant, that it is ſeveral inches thick between the ſkin and the fleſh. They are full of blood, which, when they are deeply wounded, guſhes out with as much force as when you open a vein in a fat perſon.

The animal which Meſſ. Donat and le Roy ſaw when they went on ſhore for greens, was probably a ſea-lion, of which I ſhall ſpeak afterwards; though they deſcribed it with hanging ears, as long in proportion as thoſe of a ſpaniel.

Such is the form and figure of the ſea-wolves which we ſaw in ſome little iſlands in the bay where we anchored. Thoſe whoſe deſcription and ſhape admiral Anſon has given are ſea-wolves of the larger kind. He calls them ſea-lions improperly for the reaſon hereafter aſſigned. See the plate.

Theſe animals are all amphibious, and moſt commonly paſs the night and part of the day on ſhore. When you pierce the thickets of corn-flags, in which they make their retreats and where they form a kind of apartment, you almoſt always find them lying aſleep on the dry leaves of thoſe plants. When they are in the water, they every now and then raiſe their head and part of their neck above the ſurface, and remain ſome time in this poſition, as if attentive to what is going forward. They make a noiſe much reſembling the roaring of a lion: the young ones ſeem to utter a hollow ſound, ſometimes bleating like ſheep, ſometimes lowing like calves. The larger and the ſmaller kinds move heavily, and ſeem rather to drag themſelves along than to walk, but with as much expedition as their bulk will allow. They live upon graſs, fiſh, and other animals when they come in their way. On the little iſland where our gentlemen killed ſo

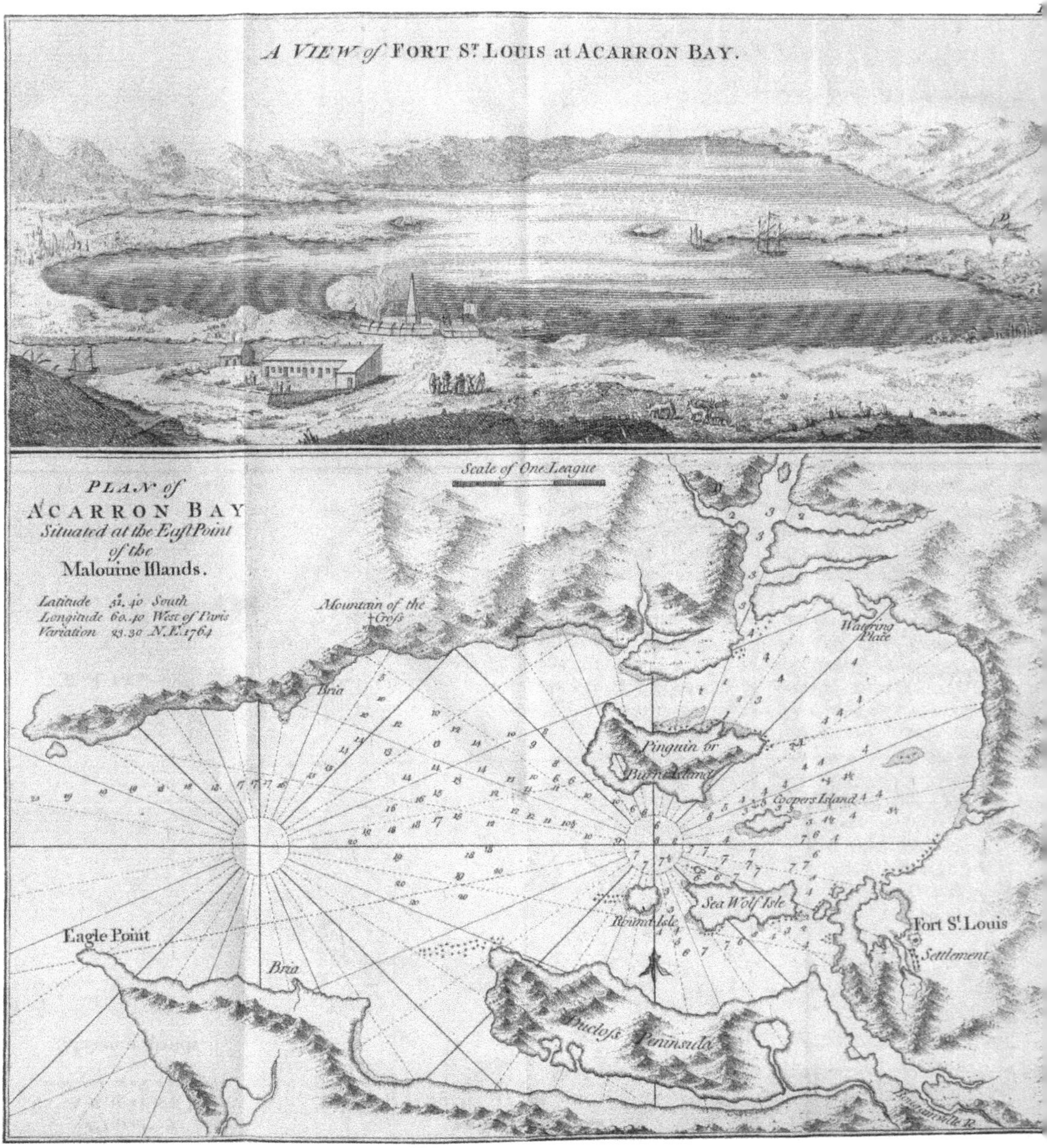

The material originally positioned here is too large for reproduction in this reissue. A PDF can be downloaded from the web address given on page iv of this book, by clicking on 'Resources Available'.

ſo many of theſe animals, one of the females ſeized a penguin at the inſtant it fell by a muſket-ſhot. The ſea ſhe-wolf carried it into the water, and devoured it ſo entirely in a moment that nothing remained but a ſlip of ſkin floating on the ſurface. M. le Roy had, the day before, brought one of theſe penguins on board, which was at leaſt two feet and a half high. We ſhall give the deſcription and repreſentation of this hereafter.

On the ſame day, while ſome of the company were employed in ſhooting, M. Duclos our captain and M. Chenard de la Gyraudais went to the top of a kind of hill toward the South, where they planted a croſs of wood about three feet high on the ſummit between two rocks, and called this eminence the *mountain of the croſs.*

On Sunday morning, the 5th inſtant, the weather being pretty fine, with a tolerable breeze, the longboat was ſent on ſhore, to get hay and water, which we did conveniently on the ſouthern coaſt, to the S. S. W. of the ſhip, where there ſeemed to be a kind of fountain. Theſe people returned on board with our ſportſmen, who brought a quantity of game of the ſame kinds as before mentioned.

About noon, M. de Bougainville and his aſſociates returned from their excurſion, much diſappointed in their hopes of finding wood. They ſet fire to the herbage of an iſland, which they have ſince called the *Burnt Iſland*, and to a promontory of the continent. They brought with them ten young penguins.

A large piece of wood which M. Duclos found on the beech gave us freſh hopes of finding ſome on the iſland. Among ſeveral others, the journals of Wood Rogers deſcribe the face of the country in the Malouine Iſlands, as conſiſting of mountains and hills covered with woods. As we had not hitherto diſcovered any in the places we had viſited, we concluded that they had only viewed it at a diſtance, and had, like ourſelves, been deceived by appearances. The difficulty however of accounting how this piece of wood ſhould be found on the ſhore, unleſs it

was

was produced on ſome coaſts of theſe iſlands, inclined us to ſuſpend our opinions till we had made ſome farther diſcoveries.

The iſland which M. de Bougainville ſet on fire, was at firſt called Penguin iſland, becauſe theſe birds were found there in ſuch numbers, that upwards of two hundred periſhed in the flames. There remained however a prodigious quantity; and we found ſome of them at every ſtep. The ſetting of this iſland on fire, which is near a full league in length, and half a league in breadth, may be ſaid to be of no conſequence, as the flames could not extend farther: but the ſame cannot be affirmed of the continent. M. de Bougainville imagined, that by deſtroying this uſeleſs herbage he was doing a piece of ſervice, as it would ſave trouble whenever theſe lands were cleared. I repreſented to him, that as the whole country was covered with the ſame herbage, the flames might probably ſpread over the whole face of the continent, unleſs their progreſs was ſtopt by ſome rivers; beſides, that they would deſtroy the game. He paid no regard to my remonſtrances; and ſet fire that very evening to ſeveral parts of the continent.

On the 6th of February in the morning, M. de Bougainville, on the credit of the journal of a Malouine captain, aſſerting that he had ſeen wood in the eaſtern part of this iſland, determined to purſue his diſcoveries. As ſoon as we had put to ſea with our cutter and longboat, the wind began to blow with ſome violence from the W. S. W. which determined us to poſtpone our expedition; eſpecially as M. de la Gyraudais propoſed to diſpatch M. de St. Simon, with three or four other officers belonging to his veſſel, to make diſcoveries by land; which was agreed to.

M. de St. Simon, a lieutenant of foot, who had lived many years with the ſavages of Canada, young, hardy and enterprizing, and in every reſpect qualified for an expedition of this nature, ſet out this very evening with Meſſ. Donat, officers of the Sphinx, and two ſeamen, to reconnoitre the N. N. W. part of this iſland.

The

The longboat of the Sphinx having likewiſe landed ſome of its crew, they found on the ſhore a bough of a dry tree fifteen or ſixteen feet in length, which confirmed us in our expectation of finding wood upon the iſland.

On Tueſday morning the 7th inſtant, the weather becoming clear, we hoiſted the anchor which had been caſt the evening before on account of the hard wind which had then ariſen. We ſent to get in ſome ballaſt and ſome water, and our ſportſmen came back about noon laden with game.

M. de la Gyraudais dined on board the Eagle, and a ſeaman brought a pretty large dry root which he found on the northern coaſt. It ſeemed to be a ſpecies of the cedar.

All theſe ſpecimens of wood determined us to make an accurate ſearch on the South Weſt coaſt. With this view M. de Bougainville, M. de Belcourt, and the Sieur Donat la Garde, lieutenant of our ſhip, embarked in the boat. They took in proviſions for three weeks, and being all well armed, directed their courſe to the South Eaſt.

On the 8th, the ſons of M. Duclos Guyot our captain, happening to throw ſome hooks at the ſtern out of the windows of the cabbin, caught a large quantity of fiſh of a moſt delicate flavour, though not more than eight or nine inches long. Their eyes were red, their gills edged with gold, and their fins of the ſame colour; their ſkin ſmooth as that of a tench. I do not know their name.

On Thurſday the 9th, at four in the morning, the wind being northerly, we got every thing in readineſs to penetrate farther into the bay. When we were under ſail the wind ſhifted to the North Weſt, which obliged us to make ſeveral tacks, ſounding all the while. We conſtantly found between twelve and fifteen fathom, the bottom of muddy ſand: at eight, the wind veering to the Weſt and blowing freſh, we anchored in a green, ſmooth, ſlimy bottom, at fifteen fathom.

On the 10th, the wind continued to blow freſh from N. to N. W. the weather was hazy, with ſhowers of rain and hail.

 We

We ſent out our longboat however to the peninſula on the N. W. of us, to ſee whether we could find paſture for our cattle. Here we killed a great deal of game. I ſaw many ſea-wolves of the leſſer kind, with a ſmooth ſkin of a dark brown. They had five claws on their fore fins which ſerved inſtead of feet, but were not divided into diſtinct toes. On our return we propoſed to ſend our cattle on ſhore, not only for the ſake of recovering them from the very feeble ſtate to which the toſſing of the veſſel had reduced them, but to free ourſelves from the neceſſity of employing a boat and men every day to procure fodder for them.

On the 11th, the wind blew too violently all the day W. S. W. to permit us to execute our deſign. At ſix in the evening, the yawl belonging to the Sphinx came on board, to inform us, that their longboat had juſt then brought back to their veſſel, M. de St. Simon and the reſt of his companions; who acquainted us afterwards, that on their return they were three days on the ſhore oppoſite to us; and had fired ſeveral times to give us notice. We heard no firing, or at leaſt none but what we thought proceeded from our ſhooting parties, which ſometimes returned very late; though always loaded with as many buſtards, teals, ducks, ſnipes, curlews, &c. as they could carry.

The gentlemen of the Sphinx farther added, that the ſuppoſed trees which we thought we diſcovered on a ſmall iſland when we paſſed near the verge of the current, were nothing more than a plant of the bullruſh kind, with flat leaves, known to our ſeamen by the name of *glajeux*: that the hillocks formed by their roots afforded a retreat for the ſea-wolves, three of which they killed as big and long as our boat, beſides ſeveral others. They likewiſe killed a kind of wild dog, much reſembling a fox of the larger ſize: ſome of the company imagined it was a grey lynx. M. Martin, lieutenant of the Sphinx, had killed two of them the ſame day.

Theſe gentlemen met with no tree; but diſcovered a large and fine bay ſome leagues diſtant from that in which we anchored.

On

On Sunday the 12th, I ſaid maſs at five in the morning, for the quicker diſpatch of the longboat intended for forage. M. l'Huillier went in the yawl to draw a plan of the bay where we anchored, and ſeveral others ſet out with him on a ſhooting party.

The wind being W. S. W. and the weather fine, the Sphinx's longboat put to ſea on Monday the 13th, with three men to make oil of the fat of the ſea-wolves, which had been killed on a ſmall iſland ſome days before. Theſe animals may with equal propriety be called porpoiſes; ſince, beſides their having fat or bacon ſeveral inches thick between their ſkin and their fleſh, they often grunt like hogs, and wallow in the ſame manner in the dirt and mire, where I have ſeen twenty of them lying down, particularly of the ſpecies deſcribed by the author of admiral Anſon's voyage, under the denomination of lions.

At the ſame time our ſmall boat was diſpatched to another neighbouring iſland in ſearch of penguins, which are as numerous as the ants in an ant-hill. Some hours after, it returned loaded with a hundred and ſixty of theſe birds without wings, ſome of which we ſalted. At ſeven in the morning we diſcovered our fiſhing-boat, at the ſouthern point of the entrance into the bay. We immediately hoiſted our colours, and the Sphinx followed our example.

At noon, the Sphinx's longboat brought on board our veſſel the Sieur Donat la Garde, and M. de Bougainville's ſervant, who we thought were ſtill in the fiſhing-boat in which they embarked. The Sieur Donat informed us, that M. de Bougainville and M. de Belcourt, had been ever ſince yeſterday afternoon on the ſouthern coaſt of the continent which encompaſſes the bay. We immediately ſent out our longboat, in which M. de Nerville, M. l'Huillier, and myſelf embarked, in queſt of them. We found them exceedingly haraſſed and fatigued, with the expedition they had juſt made on foot, through a country where there was no beaten track. We reconducted them on board, together with a ſeaman who had accompanied them. Being

 preſſed

preſſed with hunger, they as it were devoured the dinner we prepared for them, which however did not prevent them from playing their part well at ſupper, though it was ſerved up ſoon after.

They informed us, that they had traverſed the ſouth-eaſt coaſt, till they came to as fine a bay as that we now anchored in, about eight leagues diſtant by ſea, and about four by land. Here they quitted their boat, and went by land to the ſouth-weſt part of the iſland, and particularly obſerved that the coaſt ran W. N. W. which is not probable. For there is great reaſon to believe that our veſſel was then ſtationed on the eaſtern point of the iſland, the point taken notice of by ſailors in their journals; who, as well as others, were certainly deceived themſelves when they reported their having ſeen ſome large fine trees growing upon very beautiful hills. M. de Bougainville told us, that in the bay where the boat was left, he found upon the ſhore three trees which were very dry, and one of them almoſt as large as a wine hogſhead. As they met with none in all the parts of the country they had traverſed, there is reaſon to believe that theſe trees had been tranſported thither from the Terra del Fuego, or from the neighbourhood, by the waves and currents which run towards the Eaſt, the wind too uſually blowing from the S. W. and W. M. de Belcourt, M. de Bougainville's ſervant, and a ſeaman, were attacked, if it may be called ſo, by a wild dog of the ſpecies I have before mentioned. This is perhaps the only animal that is ſavage of the quadruped kind in the Malouine Iſlands: it is probable too, that it may not have been fierce, and that it only approached them out of curioſity becauſe it had never ſeen any of the human ſpecies. The birds did not avoid us, but flocked about us as if they were familiar and tame. We have not hitherto ſeen any kind of reptiles, nor any venomous animal.

The whole night of the fifteenth was rainy, and very tempeſtuous. At half paſt eleven, the thunder fell at two cables length from us, and knocked down Le Sieur Guyot our ſecond captain, who commanded the quarter deck. He received no inconvenience from this accident, except being frightened.

Our

Our longboat, which had been ſent out ſince the morning to carry proviſions to thoſe who were employed in waſhing the crews linen, could not return, on account of a contrary wind which aroſe, and blew with violence from the S. S. W.

On the 16th, about ſix in the morning, the wind fell, and the weather became hazy. Some ſqualls came on afterwards, accompanied with rain and hail. The fiſhing-boat was however ſent out to get forage. Our longboat returned about nine o'clock, and the other at three in the afternoon.

On Friday the 17th, at five in the morning, the wind blowing freſh from the South South-Eaſt, Meſſrs. de Bougainville, de Nerville, de Belcourt, Donat, de la Garde and myſelf, embarked in the cutter, with a tent and bedding, to eſtabliſh a ſettlement on the land, and to form a camp on a ſmall eminence almoſt at the bottom of the bay.

As ſoon as we landed, we ſet about pitching our tent on a ſpot which we judged to be the moſt commodious, at the diſtance of a muſket ſhot from the ſea. The little hill ran from Eaſt to Weſt. The place where we fixed our eſtabliſhment is open to the North, which makes the South of the country with reſpect to the equator. Below us, at about a piſtol ſhot from the tent, ran a rivulet of ſweet water very palatable to drink. In the front of the tent was a ſmall eminence like that on the declivity of which the tent was pitched. Some paces from thence we dug a hole in the ground for a kitchen; where, for want of other fuel, we made uſe of broom. We likewiſe tried the large green tufts of the reſinous gum-ſhrub I have mentioned. They keep up and continue a fire extremely well; but when green are not proper for dreſſing victuals.

As I ſaw the inconvenience attending the want of wood in a country where we intended to eſtabliſh a colony, I endeavoured to hit upon ſome expedient to obviate it, at leaſt till the government could take meaſures for diſpatching ſome pinks and ſchooners to be ſtationed in this country, and to make voyages to the Terra del Fuego, to bring wood for fuel as well as for build-

ing

ing and carpenter's work. I thought we might poſſibly find ſome coal, or at leaſt turf. Accordingly I equipped myſelf with a mattock, and proceeded on my ſearch. Having obſerved, that the banks of the rivulet were rather marſhy, I conceived, that, as the country had never been cultivated, the graſs which grew there, might in proceſs of time have formed a maſs of earth intermixed with roots and decayed leaves, which would exactly furniſh us with the ſort of turf I was in queſt of. In fact, after a few ſtrokes with the mattock, I diſcovered a turf of a reddiſh caſt, which was owing to its not being arrived at the maturity requiſite to give it perfection. When I had gone twenty paces up the rivulet, and had found, on digging, ſome turf with the properties I wanted, I carried two or three ſquares of it to M. de Bougainville, and acquainted him with the diſcovery. He was ſo anxious leſt it ſhould not prove the right kind of turf, that he declared it his opinion that it was not. It was ſhewn to every body who landed with us, and thoſe who were acquainted with turf, were of my opinion. M. de Bougainville ſtill in ſuſpence, wiſhing that it might be the true ſort, and yet fearing the contrary, reſolved to make a trial of it. Some dozen of theſe ſquares were dug up and ranged round the fire. Our impatience prompted us to throw a few into the fire, when we had the ſatisfaction to find, that as ſoon as the moiſture of the turf was exhaled, it burnt as well as the beſt turf produced in France and other countries. We then ſent three or four ſeamen to cut a quantity, and to pile it in the uſual manner to dry, and be ready for any uſe we might think proper to make of it.

When ſome piles of this turf were raiſed, the Sieur Donat recollected that he had ſeen in company with M. l'Huillier along the coaſt, a black fibrous earth which was tolerably dry, and might anſwer the ſame purpoſe. But having forgot the place, Meſſrs. de Bougainville, de Nerville, l'Huillier and myſelf, went in ſearch of it that day, but without ſucceſs.

While we were thus engaged in forming our ſettlement, meaſures were taken on board to penetrate farther into the bay, with a view both to be nearer us, and to provide for the ſecurity of our frigates.

Accordingly, as ſoon as we were gone, the two veſſels ſet ſail, and by proper manœuvres came at laſt to anchor immediately under Penguin's iſland, or the Burnt iſland, and within the narrow channel or entrance which one muſt paſs to get into the creek, on the borders of which we had fixed our eſtabliſhment. Near the place of anchorage there is a ſmall iſland, which has ſince been called Cooper's iſland; our people having reſorted thither to repair the caſks belonging to the ſhip.

On Saturday the 17th, in the morning, we put into the great boat the two Acadian families we had brought with us to make a ſettlement on this iſland, and to people it. At nine in the morning they landed with all their clothes, furniture and neceſſary utenſils, proviſions, and ſome tents to accommodate ſuch of the crew as were to remain on ſhore to aſſiſt in eſtabliſhing the ſettlement.

Marks of the new anchorage. The northern point of the burnt iſland, which concealed the mouth of the bay from our ſight, bore Eaſt North Eaſt, three degrees North. The center of the round iſland North Eaſt, three degrees Eaſt. The eaſtern point of the iſland abreaſt of us, N. N. E. five degrees Eaſt. The higheſt mountain at the bottom of the bay, South Weſt, five degrees Weſt. The Sphinx was anchored about a cable and a half length nearer the mouth of the bay, than our frigate.

Till this time eight of us, Meſſrs. de Bougainville, de Nerville, de Belcourt, l'Huillier, Donat, and myſelf, with two ſervants belonging to Meſſrs. de Bougainville and Nerville, had lain in one tent. We placed our matreſſes upon hay and broom, to ſecure ourſelves from the damp. Though we were very much crouded, eleven of us lay there on the night from the 18th to the 19th, our company being enlarged by the arrival of M. de St. Simon, Lieutenant of foot, Mr. Balé, ſecond Surgeon, and a Pilot, who were not provided with a tent to ſleep in.

On

On Sunday the 19th we landed a great quantity of provifions and tents, and every one difpofed of himfelf as well as he could. We had notwithftanding no fewer than twelve in our tent on Sunday night. Finding ourfelves fo much crouded we refolved the next day to pitch fome additional tents, and to feparate from each other. I was the only perfon that remained with Meffrs. de Bougainville and Nerville.

While fome were employed in fixing the tents, others went a fhooting, and returned laden with game of the feveral forts I have defcribed. M. de Bougainville, having in purfuit of game wandered a little way from his companions, difcovered another creek, formed by the fame bay, near three quarters of a league from our encampment. All along the banks of it he found a lamellated earth of a brown colour almoft approaching to black, which was undoubtedly the fame which Meffrs. l'Huillier and Donat had feen fome days before. M. de Bougainville having fhewn me a piece he had brought from the place, I pronounced it excellent for the fame purpofes as turf. We made a trial of it, and it fucceeded extremely well. Thofe who intended to ftay on thefe iflands, with a view of eftablifhing a new colony, were tranfported with joy at the difcovery, efpecially as this turf is at prefent dry and ready for burning, and as, according to M. de Bougainville's accouut, it is found in fuch abundance that the boats may be loaded with it every day and brought to the fettlement.

Walking along the coaft in the afternoon, I gathered feveral fhells, Patellæ, Cochleæ, Magellanic mufcles, &c. among the roots of that fea-grafs, called by our feamen Baudreu, which had been lately thrown on fhore from the bottom of the fea.

We this day landed the horfes, calves, cows, fheep and hogs, which we had taken on board at Montevideo. They were all fo haraffed by fatigue and ficknefs that a mare and her foal died on the beach a few hours after they were fet on fhore.

On the 21ft we were much afraid that we fhould not be able to fave any of our horfes, cows or fheep, confidering their miſerable and weak ſtate when they were landed; as they all ſeemed to be either lame or languiſhing. We left them on ſhore to take their chance, and thoſe which could not ſtand upon their feet, we dragged upon the graſs, which was at a little diſtance. Having ſent ſome perſons this morning to ſee whether they were dead or alive, they were ſurpriſed to find neither horſes nor ſheep, and the cows and calves diſperſed about the country. They were unable to conceive, that, conſidering their ſickly ſtate the evening before, they could in one night have acquired ſufficient ſtrength to run about the fields: and it was apprehended that they might be devoured by the ſea-wolves, or ſome wild beaſts unknown to us; but the carcaſſes of the mare and foal which ſtill remained on the beach, removed this ſuſpicion.

Since Sunday afternoon we were employed in chuſing a proper place for building an apartment for the reception of the parties who were to remain on this iſland. The ſame eminence on which the tents were pitched was judged to be the moſt convenient. M. l'Huillier, Engineer and Geographer to the King, marked out the foundation, according to a plan he had communicated to Meſſ. de Bougainville and de Nerville. From the Monday morning every perſon on ſhore took the mattock or the ſpade to dig the foundation.

I had ſeen the firſt plan; ſeveral alterations in which having been made in conſequence of my remonſtrances, I thought myſelf equally at liberty to give my opinion on the choice of the ground. I obſerved that in heavy rains, or when the ſnow melted, the great quantity of water which would come down from the hill would overflow the building, and if it did not inſtantly demoliſh it, would at length effect its ruin by ſapping the foundation; the declivity being rather ſteep in this place. M. l'Huillier propoſed to obviate this inconvenience by cutting a trench above to receive and carry off the water; but this did not appear to me a ſufficient expedient, as the trench could not ſtop the impetuoſity of the torrent; beſides that the water which would be detained in it, by gradually oozing through the earth, would cauſe a dampneſs in the apartments very prejudicial to the health of the inhabitants, their proviſions and furniture. My opinion ſeemed at firſt to be diſregarded: M. l'Huillier defended his own, and had already cauſed ſome lands to be cleared on the ſpot to which he gave the preference. But on mature deliberation he fixed upon another ſituation on the ſame hill, at a muſket-ſhot diſtance, where there was a very gentle declivity. The workmen were immediately ſet to dig the foundations. The ſailors belonging to the two frigates were employed in this ſervice; M. de Bougainville paying them for their day's work, excluſive of their ſeamen's wages.

On Wedneſday the 22d there were only ten men left on board the Eagle; all the reſt were employed in the building.

On the 23d ſome proviſions and utenſils were brought on ſhore from the veſſel; and our ſportſmen furniſhed an ample ſupply for the ſubſiſtence of both the ſhips companies.

M. de St. Simon, one of the keeneſt of our ſportſmen, meeting with a ſea-wolf larger than any we had yet ſeen, near the creek where we diſcovered the turf, killed it inſtantly by a lucky ſhot. On his return he related his adventure at ſupper, aſſuring us that this ſea-wolf was ſo thick and long that our boat could not contain it. Every body thought the account exaggerated. But from the deſcription he gave of its figure, I began to think that it might probably be of the ſpecies mentioned in Admiral Anſon's voyage by the name of *ſea-lions.*

Full of this idea, and being curious to know the truth of the matter, I determined to go to the place the next day, being the 24th, with M. de St. Simon and two others.

When we came within ſomething more than a thouſand yards diſtance of this animal, it appeared like a ſmall hill, riſing from the level of the ground where it lay. M. de St. Simon added to the deception of our ſight, by pointing out this pretended hill, telling us that the animal lay dead near it; ſo that we did not obſerve the ſea-wolf till we were near enough to ſee it diſtinctly. On meaſuring we found it nineteen feet and ſome inches long.

We could not at that time meaſure its bulk, being unable to raiſe or turn it in order to paſs a cord round it.

After we had thoroughly examined it, M. de St. Simon led us to the borders of another creek, thirty paces from this ſpot, where there was a great quantity of cornflags. On coming to the place, he fired at a ſea-wolf, no bigger than a very large calf, and killed it. We immediately heard on all ſides, from among theſe cornflags, cries reſembling the grunting of hogs, the bellowing of bulls, the roaring of lions, ſucceeded by a ſound like the blowing of the largeſt pipes of an organ. We could not help being rather alarmed; but recollecting immediately that theſe different cries muſt proceed from theſe animals, and knowing that we might approach them without danger, taking care only to keep off about the diſtance of their length; we entered among theſe cornflags. M. de St. Simon fired at a ſea-wolf which was neareſt to him. The ſhot entered an inch above his eye, the animal fell under the ſtroke, and died almoſt inſtantly. A fountain of blood iſſued from the orifice, and ſpouted to the diſtance of at leaſt half a foot. More than thirty pints ran out in leſs than half a quarter of an hour.

Thirty of theſe large ſea-wolves were lying two and ſometimes three in the ſame hole or pit, full of mud and dirt, where they wallowed like hogs. M. de St. Simon ſingled out ſuch as lay on dry ground, as it was more eaſy to remove them when dead, and leſs troubleſome to ſkin them, in order to get their greaſe or lard for making oil. He killed eleven of them ſucceſſively. Two others, rather larger than the reſt, being only wounded, though they had already loſt twenty pints of blood, had ſtrength enough left to get out of their holes, and eſcaped to ſea, where we ſoon loſt ſight of them. The reſt which were not wounded remained quietly in their retreats, without ſhewing any ſigns of fear or rage. Only one of thoſe which were mortally wounded, in his laſt ſtruggles ſeized ſome of the cornflags that ſurrounded him, tore them in pieces with his teeth, and ſcattered them about; but without bellowing or making any noiſe.

An Acadian who accompanied us ſkinned a young ſea-wolf, the firſt that was killed, as well as two other ſmall ones which were killed after the largeſt. Theſe are of the ſame ſpecies with that which we took for a hillock. They are exactly the ſame monſtrous animals, as are deſcribed by the author of admiral Anſon's voyage, under the article of the iſland of Juan Fernandes, ſituated at a ſmall diſtance from the continent of Chili. The whole of his relation is pretty near the truth, except that in theſe ſea-wolves, which he calls *lions*, the two feet are furniſhed with toes having diſtinct articulations, but connected by a membrane or black pellicle, and that theſe toes are armed with claws; a circumſtance wanting in the figure inſerted in the 100th page of that admiral's voyage.

The leaſt of theſe large ſea-wolves which were killed by M. de St. Simon, was from fifteen to ſixteen feet in length.

When they ſee any one approach them, they uſually raiſe themſelves upon their paws or fins as deſcribed in the plate. They open their mouth wide enough to admit eaſily a ball of a foot diameter; and keep it open in this manner, at the ſame time filling a kind of trunk they have upon their noſtrils with wind. This trunk is formed by the ſkin of the noſe itſelf; which ſubſides and remains empty when they ceaſe to bellow, or do not fill it with their breath. Their head is ſhaped like that of a ſhe-lion without ears.

Among the numbers that were killed, I obſerved ſeveral which had no trunk, the ſkin of their noſe had no wrinkles, and their ſnout ended rather in a ſharper point. Perhaps theſe were the females. All thoſe we ſkinned were males: but ſix were left lying upon their bellies in the mire without being turned; and theſe were juſt the number we ſaw without trunks. If theſe were really the females, there ſhould be much leſs difference in ſize between them and the males than is repreſented by the author

thor of the voyage juſt now quoted; for the difference is not even apparent.

While theſe animals kept their mouths open, two young people diverted themſelves with throwing large ſtones into them, which they ſwallowed as we would a ſtrawberry. They move their bodies with ſome difficulty, but can turn their head and neck to the right or left with tolerable agility conſidering their bulk. It would be dangerous to come within their reach; as they could bite a man in two with a ſingle bite. They have the fineſt eyes imaginable, and there is no fierceneſs in their countenance: I remarked that when they were expiring their eyes changed colour, and their cryſtalline lens became of an admirable green. Some of theſe animals were white, others tawny; the major part of the colour of the beaver, and ſome of a light fawn colour.

On Saturday the 25th, M. de Bougainville propoſed at breakfaſt to both land and ſea officers, to undertake the erecting of a fort upon the riſing ground forming the hill, on which the habitation or place of reſidence was built for the coloniſts, who were to remain on the iſland. We all unanimouſly agreed to erect it with our own hands, and to complete it without the aſſiſtance of the reſt of the ſhip's company.

As ſoon as breakfaſt was over, M. l'Huillier and M. de Bougainville went to chooſe the ground, and M. l'Huillier aſſiſted by two pilots marked it out upon the ſpot.

In the mean time ſome perſons were diſpatched in ſearch of tools for the execution of our deſign; others went a ſhooting to procure proviſions for the company. We had hitherto killed more game than was ſufficient for the ſubſiſtence of the crews belonging to the two frigates. We had more than once conſidered it as a ſingular circumſtance that we ſhould come with an intent to form a ſettlement in a deſert and unknown country, having no other proviſion than bread, wine, and brandy; and yet free from any care for the next day, in full confidence that the game we met with would furniſh a ſufficient ſubſiſtence for

above

above a hundred and twenty perſons, who had landed and were encamped under the tent. So far from experiencing any want hitherto, we had made ſo plentiful a proviſion, that there was no probability of our being reduced during the ſtay we propoſed to make. Nevertheleſs each meſs, conſiſting of ſeven perſons, was allowed one buſtard and a gooſe, or one gooſe and two ducks, or two geeſe, or two buſtards and ſome diving water-fowl, which we call *Becſics*, or *Nigauts*, and which I ſhall ſpeak of in the ſequel.

About three in the afternoon, we met at the place where the fort was marked out, which we agreed to call *Fort du Roy*, or Fort Royal. Every body ſet to work with ſo much cheerfulneſs, and ſuch incredible ardor, that we had the very ſame evening dug part of the ditch ſix feet broad and one deep. M. de Bougainville's example animated us all.

On Sunday the 26th, both the ſhips companies aſſembled at the habitation to hear maſs. There remained on board the Eagle only three men and two officers, one of whom had received a hurt in his leg. They all dined on ſhore, and the boat did not return to the ſhip till evening, when the wind which had all day blown with ſome violence, was abated.

On Monday and Tueſday, the longboats took in ballaſt for the Sphinx. Some poultry, beams, planks, &c. were carried on ſhore. The works were continued at the building and the fort. In my walks, I now and then took notice of the ſoil of the adjacent country. I found a pretty large quantity of ſpar and quartz; which is an indication of mines. I likewiſe met with ſome earth of a reddiſh caſt, reſembling oker, and ſome ſtones of a ruſty colour and very ferrugineous, which I ſhewed to M. de Bougainville.

I am perſuaded that there are mines of different ores in this iſland: I broke a piece of ſpar mixed with quartz with an iron crow; and perceived in the crevices a greeniſh ſubſtance which appeared to me like verdigreaſe. On touching it with my tongue the

the taſte and ſtyptic quality of this mineral was ſo ſtrong, that it made me ſpit for a full quarter of an hour.

On Thurſday the firſt of March the weather which was hazy, with ſqualls of wind and ſome rain, retarded our works; but we got ballaſt for the two frigates. M. de Bougainville came to a reſolution that the Sphinx on her return home ſhould touch at Guadeloupe, to diſpoſe of ſome of the merchandize we had on board; and that our frigate, after we had made ſome farther diſcoveries of the land, ſhould return to France, inſtead of going to the iſland Mauritius, which was intended if the Malouine Iſlands had not been fit for eſtabliſhing a commodious and advantageous ſettlement.

On the ſecond of March, at nine in the morning, we landed four pieces of cannon out of the ten which the Eagle was to furniſh for the defence of the fort we were erecting. Four more will be added from on board the Sphinx; two braſs field pieces, which were bought at St. Malo's two days before our departure, and ſix pedereroes.

As we had determined to raiſe a pyramid in form of an obeliſk in the center of the fort, I propoſed to place a buſt of Lewis the fifteenth upon the top, and undertook to execute it in terra cotta. I had ſeen ſome grey-coloured earth on the banks of a creek, which I thought very fit for this purpoſe. At ten o'clock I ſet out with our captain M. Duclos, to ſearch for it, and to obſerve what progreſs was made in extracting oil from the greaſe or lard of the large ſea-wolves, which we had killed ſeveral days ago, and left upon the ſpot. We went thither in the fiſhing-boat.

Though they had been killed ſo long and were expoſed to the heat of the ſun, which had melted a great part of the fat, the people employed in extracting this oil, aſſured us, that every ſea-wolf yielded at leaſt two hogſheads and a half, and would have afforded more than four, if the experiment had been made ſooner.

I wanted

I wanted to get the two largeſt teeth drawn from the jaw, but it was not practicable. In breaking the jaw-bone with a hatchet, the ſtroke unfortunately fell upon the teeth ſo as to ſplit them. They are ſolid and full only towards the point: the whole of what is inſerted into the jaw-bone being hollow. I at firſt intended to have diſſected the whole head, but the enormous ſize of it obliged me to relinquiſh my deſign, on account of the difficulty attending the carriage.

I employed the remainder of the time in ſeeking ſhells among the ſea-weeds, lately thrown on ſhore by the waves. There were ſcarce any other than ſome Neritæ, with ſtripes of different colours.

The bottom of the ſhell is compoſed of the fineſt mother of pearl. I likewiſe met with ſome cochleæ and Magellanic, as well as common muſcles. Some of the laſt were between five and ſix inches long and two broad, at their greateſt diameter. At ſix in the evening we loaded the fiſhing-boat with the potter's earth and turf. Finding that it was aground, owing to its being overloaded, we lightened it to ſet it afloat. We were deceived by the ebb; becauſe the ſea, which is not very regular in theſe bays, except at the time of the new or full moon, did not riſe ſo high as we expected. It was near an hour before the boat could be ſet afloat; and that it might not be overloaded, M. Duclos and myſelf determined to return by land, and to keep along the coaſt. We marched almoſt a league over flints, ſtones, and rocks, which line this coaſt. The boatmen had orders to come to take us in at the entrance of the bay, where we told them we ſhould wait for them. We reached the place with great difficulty, the weather being hazy and the wind very high. Having waited for them three quarters of an hour in vain, and while it was very dark, we concluded that the tide, which was running down, and the high wind, which was contrary, had induced the boatmen to bear away for the veſſels. We reſolved therefore to finiſh our expedition by land, by going round the bay, which is at leaſt three quarters of a league, when we we heard the boat coming towards us. We hailed her and ſhe anſwered. After attempting in vain to put aſhore at two or three places, they came near enough at laſt to give us a fair opportunity of jumping into the boat. We intended only to croſs over to the other ſide of the mouth of the creek, and to perform the reſt of our journey along the ſhore on foot. But the ſteerſman aſſuring us that the ſea ſtill roſe, and that the tide was in our favour; perſuaded us that we ſhould eaſily get the better of the contrary wind, and that they would engage to land us in a ſhort time near our habitation. Our captain ſuffered himſelf to be prevailed upon, and we got into the mouth of the bay; but we had ſcarce rowed ten or twelve yards when the wind blew with exceſſive violence, the waves ran high, and the ebb of the ſea joined to a contrary wind was ſo troubleſome, that we could not get the better of it. Notwithſtanding all our efforts we could ſcarce proceed twenty yards. The ſea grew terrible; every wave broke with violence againſt the boat, and partly beat into it, ſo that we were already overflowed. Tired with ſtruggling in vain againſt the waves, and finding ourſelves in danger of running aground upon the ſtones which lay along the coaſt, to which the waves and the wind drove us in ſpite of all our efforts, M. Duclos ſaid we muſt return to the mouth of the bay, and there run aground. In leſs than three minutes, in ſpite of the oars and rudder, we found ourſelves driven towards the ſhore at the diſtance of about four fathoms from land. The ſea which was then extremely furious, was near daſhing the boat in pieces, and we ourſelves were in danger. Our captain told us we muſt jump into the water, and ſet the example himſelf. I followed him at the inſtant that a large wave was juſt breaking againſt the boat, and overwhelmed it entirely: the ſhock it gave made me fall into the water when I was juſt coming to the ground. I recovered myſelf ſo ſoon, that I only got wet on my left ſide, and had my boots filled with water. It was ſcarce more than two feet deep. I immediately ſteered my courſe towards our habitation, and told M. Duclos our captain,

 that

that I was going to give tidings of him, while he was engaged in getting the boat afloat in order to ſecure it. When I arrived at the habitation, I found ſeveral of our company, who were under apprehenſions on our account. Finding the weather ſo bad, ſome of them imagined that we had put ourſelves on board one of the frigates, to avoid the danger of ſtruggling againſt the wind and angry waves in a boat: others fancied that the darkneſs had compelled us to land, and that we might have loſt our way. It was near ten, and they ſtill waited ſupper for us. While I changed my clothes, the ſupper was ſerved up, and I played my part at it handſomely. M. Duclos arrived half an hour after me, and went to-bed without taking any other refreſhment than a glaſs of wine.

We imagined till now, that the creeks and the bay which formed the port of our habitation, were not well ſtocked with fiſh: that the ſea-wolves and the water-fowl, which were very numerous, deſtroyed the fiſh for food, and allowed it no time to grow large. M. de la Gyraudais yeſterday convinced us of the contrary, by bringing us ſome fiſh which made part of our ſupper. Being a ſhooting at the extremity of a creek about a league from our encampment, he came to the mouth of a ſmall river when the ſea was at ebb; where, as he told us, he caught with his hands a dozen fiſh, which were left aground upon the gravel, and were endeavouring to get back to ſea. The ſmalleſt of them was about a foot long. They were ſome of them dreſt *au courbouillon*, others fried. Every body found them excellent.

On Friday morning Meſſ. Duclos, de la Gyraudais, Baſlé, M. Duclos's youngeſt ſon, and myſelf, being deſirous of making the moſt of this diſcovery, without communicating our deſign to the reſt, got ready a net of the ſize of only three fathoms and a half, and repaired to the fiſhing place. We placed two catch nets at the ſame place, when the ſea ebbed, and caught thirty fiſh and upwards, the leaſt of which weighed near a pound and a half. We afterwards caſt a third net at the mouth of a ſmall river two hundred paces from thence, and caught a dozen of the ſame ſort of fiſh.

Encouraged by this ſucceſs, on Saturday the third inſtant we returned to our fiſhing. But the ſea having ebbed, we did not catch a ſingle fiſh. We then concluded that this fiſh came into freſh water with the tide, and went back again with the ebb. Having obſerved that numbers eſcaped through the holes of our net which was a bad one, or jumped over it, we determined to go a fiſhing with the ſeamen the next day, when the high tide occaſioned by the new moon was expected. Accordingly M. Duclos went on board, and ordered the ſean to be got ready. M. le Roy carried it in the boat to the entrance of the creek in the morning, and came to acquaint us with it. A party of us, to the number of ſixteen, ſet out immediately after dinner, with Meſſrs. de Bougainville and de Nerville at our head. On caſting the ſean only once, we took more than five hundred large fiſhes, and thouſands of others half a foot long; three-fourths of which we threw into the ſea. We kept but one ſort of the ſmall ones called by the Spaniards *Pajes*, and by our mariners *Gras dos*. This fiſh is almoſt tranſparent and of a moſt exquiſite delicacy. It is excellent when fried, and not inferior to the eel pout.

The net was ſo full, that notwithſtanding the joint efforts of ſixteen perſons, it was with the utmoſt difficulty imaginable that we dragged it on ſhore. Several fiſh jumped over it, and a great number eſcaped both at the extremities, which could not be brought together, and through the holes that were in the net. However we loaded the boat, which could not reach our encampment till the next day. The fiſh were diſtributed in great plenty for two days, among the crews belonging to the two frigates: they were eaten with variety of dreſſing; and that the reſt might not be waſted, we ſalted a barrel full of them.

This fiſh reſembles in ſhape what is called *Meuille* in Saintonge. It weighs four pounds and an half upon an average.

 The

The ſame day, juſt as ſupper was over, M. Martin lieutenant of the Sphinx came loaded with game. While he was a ſhooting, he went to diſcover the ſource of the river at the mouth of which we had caught ſo much fiſh. He informed us, that there was a vaſt bay, three or four leagues north-weſt of our encampment, of which he was not able to diſcover either the entrance or the bottom from any of the heights; that this bay appeared to him to run at leaſt eight or ten leagues within the land, and that at different diſtances he ſaw rivers and iſlands. We were charmed with this diſcovery, and reſolved to pay attention to it.

The great quantity of fiſh we had caught induced us to make a ſecond trial. On Monday the 5th, we returned to the place, but whether the fiſh had taken the alarm, or the ſea was not riſen to a proper height, we caught only ſome ſmall fiſhes and a dozen large ones.

While we were engaged in fiſhing, others went a ſhooting, and took a ſurvey of the newly diſcovered bay. As they were doubtleſs leſs fatigued than M. Martin, they found the journey not ſo long, and declared it ſhorter by two leagues. This determined M. de Bougainville and ſeveral others to go thither the Wedneſday following, being Aſh-Wedneſday.

The whole company being returned about noon, and the fort, on which the officers alone had been employed, being finiſhed, M. de Bougainville propoſed to mount the cannon which were upon their ſea-carriages at the bottom of the hill. We immediately ſet about this buſineſs. Accordingly we laid planks upon the ground, to make what is called a bridge, to prevent the wheels of the carriages from ſinking into the earth. By the mere ſtrength of our hands, without the aſſiſtance of any inſtruments or engines except crows, levers and ropes, we managed to mount one cannon, notwithſtanding the height and ſteep aſcent of the hill. When we had planted it in its proper place, it being almoſt time to conclude our day's work, we loaded and fired this cannon by way of ſignal. We then cried ſeven times *Vive le Roi!*

which

which exclamation was repeated by the workmen employed in building the apartments.

Ever ſince we ſet about building our habitation, we fired a field-piece with a pound ball, and rang a bell at five every morning, and half paſt ſeven every evening, to ſummon the men to their work, and give them notice when to leave off. At eight we rang to breakfaſt, and at one to dinner. Beſides theſe meals M. de Bougainville now and then ordered them an allowance of brandy by way of gratuity. Thus the work was actually in as great forwardneſs as if two hundred workmen had been employed.

While we were thus buſy on ſhore, the few hands which were on board the frigates were by no means idle. They landed ſomething every day for the uſe of the encampment, as ordnance, balls, proviſions, utenſils, *&c.*

On the 6th we began to ſtow our ballaſt of flints, and in the courſe of the afternoon mounted ſeven cannon in the ſame manner as the firſt. It muſt be confeſſed, that ſeamen may challenge all the world in point of dexterity in moving great weights.

When this operation was finiſhed, I ordered ſome baſkets to be filled with potter's clay mixed with argil for want of ſand proper for the purpoſe, and contrived ſo as to go on board the next day, that I might work at the King's buſt without interruption, which I found to be impracticable on ſhore, where I ſhould have been obliged to do it in our tent,. into which ſomebody was entering every quarter of an hour.

On Shrove Tueſday, at ſeven in the morning, I got into the boat in order to return on board the Eagle. I took up my quarters again in my cabbin; after dinner, I began to model the buſt in M. de Bougainville's, and being unprovided with a piece of iron to ſupport the earth upon the die, I ſupplied its place with a cylinder of wood. The head was already roughly ſketched the ſame evening.

I de-

I dedicated the 8th wholly to the finiſhing of the firſt ſketch, which was already reduced to a form. Two or three officers who ſaw it in this ſtate, encouraged me to finiſh the buſt, and I was in hopes of ſucceeding in my attempt.

Animated with this expectation, I went to work at ſix in the morning on the 9th, and was not a little diſconcerted to ſee crevices and cracks in the forehead and ſeveral other places, though the earth was very well mixed. M. Guyot and M. Baſlé coming a quarter of an hour after, were almoſt as much chagrined as myſelf, to find that the earth was not proper for the uſe I deſigned to make of it.

I aſked M. Guyot, if he had not ſeen on the coaſt a fine ſand, which when mixed with this earth might remedy its defects. They ſet out for the encampment an hour after, and gave M. de Bougainville an account of the difficulties I met with from the bad quality of this earth.

I thought I had nothing to do but to make another attempt with freſh earth mixed with ſand, but M. de Bougainville, apprehenſive that a new trial might prove abortive, determined to ſubſtitute a Flower de Luce in the room of this buſt. M. Guyot returned on board to dinner, and communicated this reſolution to me. I then deſiſted from my undertaking; and paſſed the evening upon the Burnt iſland in company with M Mauclair, where we killed ten buſtards: he had killed ſixteen the day before. While we were in queſt of game, two of our officers amuſed themſelves with fiſhing with the hook from the cabbin windows, and caught fiſh enough to furniſh a diſh for three ſucceſſive meals. The angle-rod would ſupply an equal quantity every day, if the line was but thrown one hour before the meal.

Theſe fiſh are of three kinds. The firſt reſembles a pike in ſhape, the fleſh as it were tranſparent, with a ſtripe of blue, one line in width, which runs from the gills to the tail between two yellow ſtripes. The Spaniards of Chili call them *Rovalos.* The ſecond ſpecies may be ranked in the claſs of the eel pouts, called by ſome *Loaches.* The head of that here mentioned is flat and much larger

larger than the eel pouts in France. The third ſpecies is likewiſe exquiſite, and has yellow ſtripes round the gills, as if orpiment or gum had been rubbed upon it with a pencil.

Theſe three ſorts of fiſh, which were the only ones we caught on board, are no more than between nine and ten inches long; they are uſually from ſix to ſeven. But all of them are excellent, particularly that which has the head, and nearly the figure of a pike. They bite ſo freely, that they are caught as ſoon as you throw out your line. This fiſh was one reſource, when the weather did not permit us to go a ſhooting.

On the 10th I returned to the Burnt iſland, in hopes of gathering ſome *Lépas* or Patellæ, but the ſea was too high; M. Duclos's youngeſt ſon and myſelf, after killing four wild ducks and three *Becſics*, returned on board at five o'clock.

The wild gander is of a dazzling white; its bill is ſhort and black like a buſtard's, and its feet are yellow. The bill and feet of the female reſemble thoſe of the male, but the feathers upon its back are grey. The border of the white feathers which cover the neck and breaſt is black, and forms a ſpot which takes the round ſhape of the feather. The wings of both reſemble thoſe of the buſtard; and have likewiſe a hard knob like a horn at the articulation of the pinion. After ſtripping the large feathers from the body of the female, there appeared a grey down extremely fine and very thick. The down of the male is at leaſt as beautiful as that of a ſwan. They would both make beautiful muffs *.

The teal of this country are much ſuperior in beauty to thoſe of Europe. Their bills and feet are blue, their wings green and gold, and the reſt of their bodies much more ſhining and beautiful than thoſe of the Guinea hens. I ſkinned one of them, and having preſerved the head and feet, and ſtuffed the coat with fine

* Their beauty induced ſeveral of our officers to order a great number of theſe geeſe and buſtards to be ſkinned with a view of carrying them to France; but for want of proper care, they were moſt of them loſt. Mine ſhared the ſame fate for want of room to ſtow them in my cabbin.

fine mofs, placed it in its natural attitude. I made a prefent of it to a virtuofo of St. Malo. I likewife brought to France and depofited in the cabinet of natural hiftory, in the Abbey of St. Germain des Prés at Paris, the head and feet of a large water-fowl of the carnivorous kind, which I have mentioned under the name of Quebranta-hueffos. I have given its figure, on account of the fingularity of its bill.

It would have been a defirable circumftance to have poffeffed the art of preferving the eyes of thefe animals in their natural ftate. Diamonds and rubies can by no means equal the fire, the beauty and the luftre of the eyes of a certain fpecies of water-fowl or diver, which is frequently feen on the fea-fhore.

The pupil is furrounded with a circle of the fineft vermillion or carmine. The head is black, but the feathers from the eye to the back of the head are of a fhining white mixed with fome ftreaks of black.

In thefe iflands there are likewife prodigious numbers of fmall eagles or brown hawks, of the fize of the largeft of our cocks; but the wings of which when extended, were at leaft three feet acrofs. The large feathers of the wings are of a bright yellow, mixed with brown in tranfverfe ftripes. There is likewife a kind of eagle, of the fize and colour of a turkey hen, white, red, or yellow. In this kind of eagle, at the bottom of the bill, there is a fkin of a very fine red, ftrewed with pretty long black hairs. When this bird is dead, the red colour fades, and the fkin changes to a very pale rofe colour. Its talons are fcaly and of a light grey, as well as thofe of fome of the fmaller kinds I have mentioned. The reft have yellow feet. The talons of this laft mentioned fpecies are as ftrong and large as thofe of the larger kind. Sparrow-hawks are likewife found here, with white breafts and necks; thofe of others are variegated with white, grey and red.

Mufcles are very commonly found along the coaft. We more than once attempted to eat fome of them; but found them fo full of pearls, that it was impoffible to chew them: as thefe pearls being very hard endangered the breaking of our teeth, and when they

they were broken in pieces, they left a kind of ſand in the mouth which was very diſagreeable. As I believed the production of theſe pearls to be owing to ſome diſorder in this ſhell-fiſh, I imagined that this diſorder might be owing to this animal's ſuffering from the want of water during the ebb of the ſea. I therefore fancied, that if we took ſuch as were conſtantly ſupplied with water, we ſhould find them without pearls. The muſcles I had found among the roots of the ſea graſs, confirmed me in this opinion. I opened ſome both of the common and Magellanic ſorts; they were without pearls and excellent. I carried two or three dozen to the encampment; they were liked by all the lovers of this ſhell-fiſh, and we afterwards ate them frequently.

No remarkable occurrence happened from the 11th to Thurſday the 22d of March. Proviſions and other articles were landed for the uſe of the people who ſtaid to eſtabliſh this new colony. On the 21ſt, we laid the firſt ſtone of the pyramid.

There was one circumſtance however that deſerved notice, and occaſioned various reflections among thoſe who were witneſſes of it. It was related to me on my return to the encampment.

On Thurſday the 22d of March, I was deſirous of knowing the truth of this circumſtance, and have ſince been convinced of it more than once by ocular evidence. We carried over about a dozen hogs male and female. One of theſe was caſtrated. After they were all landed they went to ſeek their livelihood in the fields, and never failed to return every evening to paſs the night together near the encampment. At firſt they had a kind of litter of hay made for them, which though in the open air they certainly enjoyed very much, as they repaired to it ſo punctually. Somebody obſerved that the caſtrated hog generally returned about half an hour ſooner than the reſt, took ſeveral turns round the litter and placed the hay in order; that he took and carried it in his teeth to their lodging, and filled every place where it was wanting. When the reſt returned they lay down together, and he took his place laſt. If any one of them found his ſituation uneaſy, he got up, and falling upon the caſtrated hog, bit him, and obliged him to fetch more hay to make up the litter. The females in particular were very nice in this article.

During our ſtay one of them brought forth eleven pigs, and another twelve. Beſides theſe young ones, we left there eight ſows and one boar. It is eaſy to judge how faſt they will multiply.

I returned to the encampment with an intention of ſtaying only three days, and ſetting out on the 23d, to go by land to a bay ſituated to the South Eaſt of the iſland. M. de Bougainville having ſeen it in the tour he made ſome days after our arrival, thought it delightful, and called it *Beau-port*, as it was well adapted for a commodious harbour. I was to accompany M. l'Huillier, and two or three others thither to take a draught of it. But as ſoon as M. de Bougainville and myſelf came to the encampment, M. l'Huillier urged the neceſſity of poſtponing the expedition to *Beau-port* to the Thurſday following, his preſence being abſolutely neceſſary to carry on the building. M. de la Gyraudais was the only perſon who returned on board the Sphinx. M. de Bougainville lay in his cott; I ſpread a mattreſs upon ſome hay in the ſame tent, and lay in this manner nine nights. I employed the day in viſiting the adjacent parts, in botanical reſearches, and in other inquiries into natural hiſtory.

On Saturday the 24th of March, it was propoſed that we ſhould go in ſearch of the three ſtray horſes, to ſecure them with ropes, and bring them to the encampment. A party of thirty who ſet out on this errand, found and ſurrounded them. They ſuffered us to approach ſo near, that M. de St. Simon ſeized one of them by the mane; but the mare which he held diſengaged herſelf by a violent effort which threw him down, and leaped with the reſt over the ropes we had put round them. They ran ſo far, that it was thought proper to give over the purſuit.

We had better ſucceſs with the cows and heifers. Theſe were in the ſame manner ſcattered and diſperſed over the country; but a little calf that had been caught being brought near the encampment, and tied to a ſtake, the dam hearing it low in the evening came to give it the teat, and the reſt followed her. By returning in this manner two or three days ſucceſſively, theſe animals became accuſtomed to it, and repaired punctually every evening to the ſtable that was built for them.

On the 27th, M. de Bougainville and M. l'Huillier ordered their cotts to be carried to the chamber in the new building, which was intended for M. de Nerville. They propoſed to me the removal of my bed, but I rather choſe to ſtay in the tent, as the damp iſſuing from the walls, which were rough-caſt this very day, might prove prejudicial.

I was near having reaſon to repent of my reſolution that very night. At ten in the evening, the wind ſprang up at South Weſt, and continued ſo violent all day with frequent ſhowers of rain, that it ſeemed as if the tent would be carried away, or blown down upon me at every blaſt. I lay there however the next night, but was obliged to change my quarters the day after, being the 29th.

The tents were ſtruck to furniſh wood for the building, I ſurrendered mine, and removed to M. de Nerville's quarter.

Meſſ. de Bougainville and de Nerville had, on the 21ſt, laid the firſt ſtone of the baſe of the pyramid, or kind of obeliſk, intended to be erected in the center of the fort. A round ſilver plate, about two inches and a half in diameter, was depoſited in the ſtone-work of the foundation; on one ſide of which was etched with aqua fortis, the draught of that part of the iſland where the fort and habitation were ſituated; on the middle, the obeliſk with theſe words for the exergue, *Tibi ſerviat ultima Thule.* On the other ſide was the following inſcription:

Difcovery.
Settlement of the *Malouine*
Iflands, fituated 51 d. 30 m. South
latit. and 60 d. 50 m. Weft long. E. of the mer.
of *Paris*, by the Eagle Frigate Captain *P.*
Duclos Guyot, Captain of a fire-fhip, and
the Sphinx Sloop Captain *F. Chênard*,
Gyraudais Lieutenant of a Frigate,
fitted out by *Lewis de Bougainville* Colonel of Foot, Captain of
the veffel, Commander of the expedition, *G. de*
Bougainville de Nerville Volunteer, and *P. Darboulin*
Adminiftrator General of the Pofts in *France*. Conftruction of
a Fort and Obelifk embellifhed with a medallion
of his Majefty *Lewis* XV. agreeable to the plans of *A.*
l'Huillier de la Serre Engineer Geographer of the
Camps and Armies ferving on this Expedition
under the Miniftry of *E. de Choifeul*,
Duke of *Stainville*. In
February 1764.
[With thefe words for the exergue, *Conamur tenues grandia.*]

This kind of medal is inclofed between two leaden plates, and the whole in a hollowed ftone. Near it is placed a double glafs bottle well ftopped with maftic to refift the wet, containing a roll of paper on which were infcribed the names, firnames, ranks, and countries of all the perfons who compofed both the fhip's companies employed on this expedition, and of the volunteers *.

This fort was called *Fort de St. Louis*. It is fituated on a rifing ground, not overlooked by the neighbouring heights which are

* This lift roll which is inferted in the original, is omitted by the Tranflator, as not being interefting to the Englifh reader. The number of perfons on board the two fhips, including officers, failors, paffengers, fervants, &c. amounted in all to 138, and 28 of thefe, including women and children, remained in the ifland for the eftablifhment of the colony.

are at the diſtance of at leaſt two full leagues. It commands all the adjacent country, and eſpecially the entrance of the creek, at the extremity of which the new habitation is built. This entrance is with good reaſon called the *Goulet*, or *Gullet*; becauſe when the ſea is high, the opening is no more than a full piſtol-ſhot in breadth.

M. Baſlé and myſelf went on the 28th to ſee the large bay, where I gathered a great quantity of the moſt beautiful *Limas*, or cochleæ, with mother of pearl, and faſciated, and ſome flat patellæ, which were extremely fine. On the 29th, it blew a ſtorm, and there fell a great deal of ſleet, attended with ſqualls of wind.

On the 30th the wind blew very cold, with hazy and dark weather, which continued all night, a circumſtance unuſual in this country, at leaſt ſince our arrival. Till this day, the 31ſt, the ſky had almoſt conſtantly been fine and ſerene. We had white froſts two or three times, and once only the ſtanding waters were ſkimmed over with ice; but for ſeveral days paſt there had been a coolneſs in the mornings and evenings, which in hot countries we ſhould call *cold*. However from ten in the morning to five in the afternoon, you feel the warmth of May in thoſe places which are ſheltered from the wind.

The weather was hazy all night, and Sunday morning the firſt of April. About ten the wind diſperſed the fog, and veered to the North North Weſt, where it blew with ſome violence, but ſubſided at four in the afternoon, when I returned on board with almoſt all the officers who were not to winter in the new colony. M. de Bougainville and M. l'Huillier, were the only perſons who ſtaid to ſuperintend the work on the roof of the building, which was almoſt finiſhed.

The ſame day Laurence Lucas, carpenter of the Sphinx, put the finiſhing hand to the carving of the double flower-de-luce in ſtone, which was to be placed on the top of the pyramid. The two medallions in wood, one repreſenting the buſt of Lewis XV. and the other the arms of France, which were to be fixed on

two

two oppoſite ſides of the pyramid were in great forwardneſs. All the proviſions and other articles which were intended to be left upon the iſland were landed, and lodged in the Magazine.

On Monday morning the 2d, M. Duclos Guyot went in the yawl to ſound the bay, round the ſmall iſland covered with corn-flags, which was the neareſt to the place where we were moored, and was called *Ile au Tonnelier*, or Cooper's iſland, becauſe our cooper was ſettled there in order to carry on his buſineſs. M. Duclos every where found a good bottom, and concluded from the depth, that the true channel of the tide is on the ſide of this iſland oppoſite to that where we were moored. The afternoon was very windy.

On Tueſday the 3d, it was calm all day, and the weather fine. M. de Bougainville repaired to the habitation, to make every preparation for taking poſſeſſion of theſe iſlands, having fixed the day for Thurſday next.

M. de Nerville and myſelf, paſſed the whole afternoon upon the Burnt iſland, where we gathered a large ſalad of creſſes and celery upon the banks of a pond at the eaſtern point. The latter of theſe plants is very common in all the parts of this iſland that we have viſited.

On the 4th, the wind which blew very freſh, varied from the South South Weſt, to the Weſt North Weſt, the weather was fine, and the ſea ran very high; which did not however prevent our ſportſmen from going out to kill buſtards. Four officers belonging to the Sphinx, had brought from thence a hundred and three ſome days before. Our officers, encouraged by this ſucceſs, determined to go thither, and killed eighty-three. Two of them killed but 36 this day, with fourteen ducks and teals. They gave 18 buſtards to the Sphinx, on account of the preparations that were making for their departure, which was fixed for the next day. The fort fired one and twenty cannon to announce the ceremony of taking poſſeſſion, which was to be performed the next day.

At

At four o'clock on Thurſday morning the fifth of April, our longboat was ſent with her hawſer and anchor, on board the Sphinx; after which ſhe weighed her two anchors, and got under ſail at half an hour paſt ſeven, with a favourable wind and fine weather.

At day-break the fort made a diſcharge of one and twenty pieces of cannon.

The moment the Sphinx ſailed, we all embarked in our yawls and one fiſhing-boat to go to the fort. As ſoon as we landed at the gullet, the fort ſaluted us with ſeveral guns. A party of the inhabitants, who had taken the reſolution to remain in this new colony appeared in arms at the gullet. They conducted us to the fort, at the foot of which we found all the reſt under arms. After the parade they accompanied us to the fort with drums beating.

All the company being aſſembled at the fort, the pyramid was opened; I then ſolemnly ſang the *Te Deum*; after that the pſalm *Exaudiat*, then thrice *Domine ſalvum fac regem*. After this I rehearſed the verſe *Fiat manus tua, Domine, ſuper virum dexteræ tuæ*; the reſponſe was, *& ſuper filium hominis quem confirmaſti tibi*, then the prayer *Quæſumus, omnipotens Deus, ut famulus tuus Ludovicus Rex noſter, &c.* for the proſperity of his reign. We cried *Vive le Roy* ſeven times and fired twenty-one cannon. We cried again ſeven times *Vive le Roy*. M. de Bougainville then produced the king's commiſſion, appointing a governor in the new colony, which was delivered to M. de Nerville, who was immediately received and acknowledged as ſuch. M. de Bougainville, in the king's name, likewiſe proclaimed the other officers, who were in the ſame manner unanimouſly acknowledged.

An altar was likewiſe erected in the fort at the very baſe of the pyramid. I intended to have ſaid maſs there, to make the ceremony of taking poſſeſſion more ſacred and ſolemn. But the wind blew with ſuch violence, that notwithſtanding a tent was erected there, it was thought proper to content ourſelves with the ceremony I have deſcribed. We afterwards repaired to the apartments

apartments in the habitation, where at eleven o'clock we had a plentiful breakfaſt, the allowance to all the ſhip's company being doubled on the occaſion.

As ſoon as breakfaſt was over, we went to viſit the ſeveral ſpots on which different ſorts of grain had been ſown eight or ten days before; and found them ſprung up, and in a very healthy and flouriſhing ſtate.

On our return, I ſtopped at a place where I had obſerved a pretty common plant, which makes an excellent infuſion: this I ſhall deſcribe hereafter. Having only time to gather a little of it, we took our leave, and returned on board.

On Friday the 6th, at ſix in the morning, M. de St. Simon and two others went on ſhore in the longboat to water, and killed ſeventy buſtards, twelve ducks, ſome teals, and ſeveral ſnipes. Theſe buſtards, together with a great number that were killed before, were put in barrels; ſo that we had two tierces and ſome barrels to ſupply us on our return to France.

The calm and the fine weather at ſun-riſe, favoured the execution of M. de Bougainville's deſign to ſurvey and take draughts of the great bay where we lay at anchor, of its creeks and the environs. With this view, Meſſ. l'Huillier, Duclos, his two ſons, Meſſ. de St. Simon, Donat, le Roy and myſelf, embarked in the longboat, and landed at the bottom of the bay in a creek, which runs up a great way within the land. You ſee it in the chart of the harbour. Meſſ. de St. Simon, Donat, and le Roy, went out a ſporting, while Meſſ. l'Huillier, Duclos, Seigneurie, ſome others and myſelf, made obſervations from the eminence or mountain E. When we had finiſhed our obſervations, and taken a draught of the bay with the graphometer, we amuſed ourſelves with obſerving a ruin, produced, as it ſhould ſeem, by ſome earthquake. It afforded a proſpect ſo dreadfully pleaſing that I was extremely mortified at my want of time, and the neceſſary inſtruments, to ſketch out a perfect repreſentation of it. A painter might here find materials to compoſe a picture of the fineſt

fineſt ruins. A ſketch of it is given in the plate, as alſo of a kind of amphitheatre ſituated a hundred paces from it.

We were no leſs aſtoniſhed at the ſight of the infinite number of ſtones of all ſizes thrown one upon another, and yet ranged as if they had been piled negligently to fill up ſome hollows. We admired with inſatiable delight the prodigious works of nature. I attempted in vain to engrave a name upon one of theſe ſtones, which formed a table a foot and an half thick, ten feet long, and ſix broad; it was ſo hard that neither my knife nor a punch could make any impreſſion upon it. I tried ſeveral in the ſame manner which were equally hard. I broke off a piece by ſtriking a corner with another ſtone, and all the pieces that were broken off had the appearance of freeſtone porphyrized.

This freeſtone as it is found in its beds, which run in all directions, is every where cut into tables of a different ſize and thickneſs; but in ſuch a manner as if art had been uſed.

Theſe ruins repreſented in the plate, reſemble in ſeveral places the gates of a city, whoſe arches are demoliſhed; and of which there remain only ſome walls to the right and left, ſtill raiſed twenty or five and twenty feet, in the parallel angles forming the entrance. They are like the walls of a town, the ſtones of which have been ranged according to the level and the perpendicular, as they are in our walls compoſed of freeſtone. Some angles are likewiſe to be ſeen here, both ſaliant and re-entrant, ſome out-works more than fifteen feet high, and ſome rectilineal projections like corniſhes, advancing at leaſt half a foot, and which run at the ſame height all along the poſterior or internal, as well as the anterior or external parts of the ruins. The only things wanting are the mouldings.

To the left of the track leading from the ſpot where we landed, we met with the eminence on which the ſtones are ranged like the arches of an amphitheatre: the figure of which I have given. Beyond theſe ruins lies a valley more than two hundred feet deep, and about half a quarter of a league

 broad,

broad, the bottom of which is covered with ſtones thrown together promiſcuouſly, and ſeems to have ſerved as a bed to a river or ſome large torrent, which running through the hollows made by theſe eminences, probably diſcharged itſelf into the great weſtern bay I have mentioned. The eminence which is beyond the valley appears to be covered with ruins, ſimilar to thoſe upon the eminence on this ſide. Before you come to theſe you meet with an eſplanade, or platform of earth, about twenty or twenty four yards broad, which runs from the baſe of the amphitheatre, beyond the firſt opening of theſe ruins, which I ſaid reſembled the entrance or gate of a city. The rubbiſh of theſe ſeeming walls obſtructs the continuation of this eſplanade where you ſee two pieces of water, or reſervoirs, one nearly round, the other oval, at a ſmall diſtance from each other; the firſt about twenty-five feet in diameter, the other thirty. A gentle declivity fifty feet broad leads from the eſplanade to the ruins.

From the bottom of the hill you ſee kinds of hollows intirely filled with theſe promiſcuous heaps of ſtones.

Between theſe hollows are irregular ſpots of ground, twelve, fifteen, twenty, and twenty-five feet in breadth, and twenty, thirty, and at leaſt fifty in length, covered with herbage and heath, as if they had eſcaped the ſhock. Between theſe promiſcuous heaps of ſtones, are every where left void ſpaces or interſtices, whoſe depth cannot be eſtimated. The ſmalleſt of theſe ſtones, none of which are angular, the corners being rounded, are two feet in length, and one in breadth or thereabouts; their figure however is not regular. They are likewiſe compoſed of a ſpecies of freeſtone which is of a very hard quality. It is an hour's walk from the place of our landing to the rubbiſh, and the road is level all the way as far as the foot of the eminence on which the ruins are ſeen.

As we returned, I gathered a little bag full of a plant which I ſhall deſcribe in the ſequel under the name of *Lucé muſqué*, or *Thé des Iſles Malouines*: I ate twenty of the fruits of a ſmall herb which

which our mariners call *Plat de bierre:* and we returned on board loaded with game.

After the ceremony of taking poſſeſſion, M. de Nerville invited us to a dinner he intended to give the Sunday following, by way of taking leave, and wiſhing us a ſpeedy return to France. We agreed to wait upon him. But our captain M. Duclos Guyot, having repreſented to M. de Bougainville, that the longer we deferred our departure, the greater would be our danger of meeting with bad weather and a tempeſtuous ſea, on account of the approach of winter in this country; that, two days ſooner or later were of conſequence, eſpecially as M. de Bougainville wiſhed to give the court as early an account of his expedition as poſſible; and that, it was therefore neceſſary to ſeize the firſt opportunity of getting under ſail: the reſolution was accordingly taken on Saturday evening, to ſail the next morning if the weather proved favourable.

On Sunday the 8th of April, we ſailed at half paſt four in the afternoon, the fort ſaluting us with twenty diſcharges of cannon. M. l'Huillier and ſome others were diſpatched early in the morning to the habitation, to make our compliments of departure, and to bring two hogs and two dozen of fowls to make broth for thoſe who might have the misfortune to be ill.

When we arrived in the great bay, that is, when we had got beyond the iſlands ſituated in it, we lay by to wait for our longboat, which arrived at ſix with our great anchor. When we had taken them and our yawl on board, we go under way at half paſt ſeven. At half paſt nine we were North and South of the iſland at the entrance of the bay. From this time to midnight, we directed our courſe to the Eaſt, at the rate of three leagues and two thirds an hour.

I could not have conceived, that at fifty-one degrees and an half latitude, and ſixty longitude from the meridian of Paris, a climate could have been found ſo temperate as that of the Malouine Iſlands. We landed at the eaſtern point, a part of the iſland expoſed perhaps more than any other to cold, white frofts, and other inconveniences incident to a fituation almoft intirely encompaffed with the fea, or with bays, forming a peninfula fwept by the South Weft and Weft winds, which are the moft frequent in thofe parts. We had reafon to draw this conclufion during more than two months ftay in the country even in the time of autumn, when the cold might be expected to be felt early in that latitude, and from the herbage in all the parts we vifited, inclining to the North Eaft and Eaft. Notwithftanding this, except the grafs which was withered by the heats of fummer, as is ufual in all other countries, the other plants, and even the grafs of the fecond growth, were ftill very green at the time of our departure.

In the quarter of the ifland which we faw, the land every where prefents a very agreeable afpect. Mountains, or rather eminences which we called mountains, encompafs plains farther than the eye can fee, divided by little rifing grounds and hills which communicate by gentle declivities. At the foot of each a rivulet, more or lefs confiderable runs in winding mazes, and difcharges itfelf into the fea through the numerous creeks of the bays. That in which we anchored (which might be called *Baye de St. Louis* on account of the fort of this name which is erected on the land which terminates it, or rather *Baye Royale*, on account of the pyramid dedicated to Louis XV. our well-beloved monarch) runs up more than fix leagues within land; and naturally forms a good harbour in which more than two thoufand fhips may ride at anchor. There is every where a good bottom, iflands of different fizes, peninfulas to the number of about twelve, which afford fuch fhelter from the moft violent winds, that perhaps there is never any fwell in thofe parts.

The entrance of this bay is at leaft two leagues over, and is contracted by a pretty large ifland at fome diftance from the South Eaft point, as may be feen in the chart.

This great bay which was difcovered fifteen days before we left the ifland, has been examined and traced in part by M. de Belcourt and M. Martin, who made an excurfion thither of two

or

or three days and nights. As we were desirous of obtaining a more perfect knowledge of its extent, Mess. de St. Simon and Donat set out some days after the return of the two gentlemen I have just now mentioned. They went at first to the place where it approaches nearest to the habitation, which is at the distance of two little leagues, and then kept along the shore till they came to the bottom of it. They passed here to the opposite shore, and followed it ten leagues. The brooks and a considerable river which it was difficult to cross, obstructing their farther progress, they determined to climb the highest mountain they could find; from whence they thought they should be able to discover the entrance of this bay and the rest of its course. They judged at that time that it ran at least fifteen leagues within the land, and formed into a peninsula that part of the country where we had established our settlement.

According to their account, the coast of this bay presents to the view an excellent soil, and an agreeable prospect. At every quarter of a league it is watered by brooks and small rivers, one of which, that runs from the West, appeared to them to be sixty feet broad. They found a prodigious number of bustards in flocks of twenty or forty, and a great many other birds. Upon the whole, they counted twenty-six pretty large islands in that part of the bay which they surveyed.

It may be doubted, whether there is not actually a streight which divides these islands, and communicates from North to South as some navigators have imagined, and whether the appearance of this bay might not have led them to form such a conjecture. Perhaps they saw only its entrance, or not venturing on account of its running so deep within the land and its great breadth, to proceed farther into it, concluded that it formed a streight *

After an attentive examination of the soil at the habitation, and that of its environs, I think I may venture to pronounce it of

* It was found on a second voyage, that such a streight actually exists; and that its entrance on the northern side, is at the place called by us *la Conchée*.

of a mineral nature. The ochreous earths, both red and yellow, the ſpars, the quartz, which are every where to be found, are evident proofs of it. The rocks which are commonly covered with grey and reddiſh ſlate, ſufficiently indicate a great quantity of ſulphur. On breaking the tops of the rocks of quartz which appear on the ſurface, with crows and mattocks, I found in the crevices, marks of a vitriolick and coppery matrix. I likewiſe diſcovered a ſubſtance of a greeniſh caſt, which had the aſtringency and acidity of verdegreaſe: I applied a little of it to the tip of my tongue, and was forced to ſpit very much for a full quarter of an hour. Here you frequently meet with pyrites which are round, and ſulphureous; and with others of irregular figures, which one would conclude belonged to an iron-mine, both on account of their weight and their brown colour, mixed with an ochreous earth of a reddiſh yellow, or of the colour of ruſt. In digging to lay the foundations of the houſes, M. de Bougainville obſerved in the earth that was thrown up, ſeveral pieces of broken quartz, which exhibited to the eye ſpangles that glittered like gold. He picked up ſome which he brought to me, and I imagined at firſt ſight that it might be *mica*, or the yellow talc. However as the talc is not uſually found in the quartz, I thought it might be that ſpecies of ſulphur which glitters in the pyrites. We were unfortunately unprovided with the neceſſary materials for making experiments; we had no coals, or wood, no furnace, or even aqua regia, nor could any be made with ſo ſmall a quantity of aqua fortis. The crucibles I brought were uſeleſs to me. There was beſides, too ſmall a quantity of theſe little glittering particles, and we had too many other objects to engage our attention, to be at leiſure to ranſack the earth for ſuch a collection of them, as would be ſufficient to make an experiment. I therefore contented myſelf with viſiting the place where the ground had been dug, and examining the earths that were thrown up. In a hollow at the depth of about ſix feet, I perceived a bed of earth lying obliquely, ſix inches broad in ſome places, the reſt of an unequal breadth, which entered the ground in the ſame direction. This bed was compoſed of quartz covered with a ruſty earth, yellow and red ochre, and a ſort of hollow flints, ſeveral of which were filled with a ſpecies of fine bole, of a fleſh or roſe colour in ſome, and of the colour of fine lacca in others; ſeveral, with a very fine earth of a browniſh red colour. The cover, or ſtony cruſt which ſurrounds theſe fine earths, is commonly of the ſame colour with the incloſed ſubſtance. I have met with ſome of them grey, very much reſembling ſilver ore. Their colour became rather deeper when expoſed to the fire, which gave me reaſon to conclude that they are of an ochreous quality, and that they conſiſt chiefly of iron. On my return to France, I ſhewed ſome of theſe pieces of quartz to perſons verſed in experiments on foſſils, who likewiſe judged them to be iron ore.

Having therefore no hopes of making diſcoveries of this kind, I turned my attention on the plants of the country. I met with only four or five of thoſe kinds which grow in France. Here is plenty of red and white celery, which has a ſweet and pleaſant taſte, though produced without culture We ate it in ſallads and ſoups every day. Some of our mariners called it *Macedonian Parſley,* and were afraid of it at firſt, but ate it afterwards without ſcruple, eſpecially as the country afforded no other greens.

M. Duclos, captain of the Eagle, found ſome hartſhorn, or rocket which he called *Creſſonette,* and brought it to the encampment. On taſting it, we found it rather too poignant. As M. de Nerville and myſelf were walking along the ſide of a pond, we met with ſome very good creſſes, and frequently ate them mixed with celery. Along the banks of a little rivulet, I have ſeen the *Grenouillette* or crowfoot, as well as the ranunculus, which is cultivated in gardens for the beauty of its flowers.

Our pilots obſerving that we were fond of a plant which had rather a ſingular appearance, were induced to taſte it. It has a milder and more agreeable acidity than even the round leaved ſorrel. They found it ſo palatable that they put it into their ſoup

ſoup the ſame day, and as we did not perceive that they ſuffered any inconvenience from it, we ordered ſome of it in our own.

This plant produces leaves ranged in a circular form, ſometimes eighteen or twenty in number, at the extremity of a cherry-coloured petiole as thick as a crow quill, round and generally from ſeven to eight inches high, always riſing above the plants which ſurround it. The leaf is of a light green.

It has only one ſtem nearly ſimilar to the footſtalk of the leaves, which ſupports a ſingle white flower conſiſting of a pentaphyllus calix, and having the figure of a very ſmall tulip; it expands in the ſame manner, and emits a very ſweet ſmell like the almond. The leaf of the plant is ſhaped like a heart, the extremity of which is very much lengthened: each leaf is faſtened to the petiole or footſtalk by this extremity, and forms a kind of hoop. See the figure in the plate. I have never ſeen any of theſe leaves quite expanded; they are almoſt always ſunk into a channel. Ten, twelve, and often more of theſe leaves or leafy ſtems proceed from the point of a long twiſted root, covered with ſmall pointed ſcales of a red colour inclining to vermillion, lying horizontally two or three fingers deep. This plant is very common. We called it *Vinaigrette* from its taſte. Perhaps it belongs to the claſs of wild ſorrel.

The plant repreſented in the figure next to that of the *Vinaigrette*, may be ranked among the *Satyrions:* its leaf ſeems at firſt ſight to ſuggeſt this conjecture: however as the orchiſes have uſually no more than two tubercles at their root, and this plant has twelve roots or more, ſhaped like thoſe of the goat's beard and very long, I think it ought not to be placed in the claſs of the orchiſes. I take it to be the Epipactis, mentioned by Father Feuillée, page 729. pl. 29. under the denomination of *Epipactis amplo flore luteo* vulgo *gravilla:* the root of the Epipactis of the Malouine Iſlands, bears however a greater reſemblance to that of the *Epipactis floribus uno verſu diſpoſitis* vulgo *Nuil*, which he mentions p. 726. and is repreſented in pl. 17. It grows in the dry and barren parts of Chili, and the Epipactis

flore

flore luteo in the moiſt parts of the ſame country: that of the Malouine Iſlands likewiſe grows in low and moiſt places. The root of this conſiſts of ſeveral knobs, which are formed into a bunch. I have ſeen from ten to twelve of them, and ſometimes more. Their length on an average is three inches, and ſome of them are more than half an inch in thickneſs. They are covered with a ſmall thin ſkin, incloſing a friable, ſoft, watery ſubſtance, which at firſt has a ſweetiſh taſte, but when it is chewed leaves ſo ſtrong a flavour or reliſh of ambergreaſe in the mouth, that it a little reſembles cat's urine.

I have not been able to diſcover the flowers of this plant, though I have ſeen ſeveral of every ſize. The higheſt of them have capſules filled with ſeed, and a kind of tuft at the extremity reſembling a cluſter of dried petals of a reddiſh caſt, without any determinate ſmell.

The ſeed is a very fine red duſt, that fills the hollow part of the capſule, which is divided into four or five compartments. After the moſt careful examination, I have not been able to diſcover any other kind of ſeed.

The ſtem of the plant never riſes higher than ſeven or eight inches, and is covered with pretty long leaves, which frequently form a ſhallow channel; ſome are perfectly flat: they are all ſmooth, and of a green colour, reſembling that of the leaf of the orchis.

In all places waſhed by the water there is found a ſpecies of ſpleen-wort, which grows like a fungus, and ſupports a ſtem with hollow leaves, in which the ſeed is contained: a circumſtance not common in any ſpecies of the maidenhair, in which the ſeed is a duſt adhering to the verge of the prone diſk of the leaf. In this plant we are deſcribing, the ſtem that ſupports the ſeed riſes ſingly on the right ſide of the root, the leaves being at the ſame time circularly diſpoſed, or vertical. The ſtem itſelf, or if you will, the only leaf in the whole plant, which ſupports the ſeed, is in proportion near an inch longer than the longeſt of thofe leaves which grow out of the fame root. This feed however, like that of the maidenhair, is a thick red duft.

In the fields, amongft the herbage which covers almoft the whole furface of the foil of the ifland, there is a pretty common plant with a white flower, radiated like that of the dandelion, but the petals are fharp pointed. The leaves the largeft of which are three inches long, and the ftem, which is about a foot high, are of a green colour and rather foft like cotton. Upon each ftem is a fingle flower, which fmells exactly like Benzoin.

There is another plant, whofe ftem and leaves refemble thofe of the preceding, which bears a bunch of yellow flowers, twelve or fifteen in number, equally radiated, and very pleafing both to the fight and fmell. This flower is fupported by a fquamous calyx. The root is a mafs of fmall fibres, all terminating at the bottom of the plant.

Here are alfo two plants to be met with, which both produce a red fruit: the fruit of one of them fo much refembles a rafberry, that it is eafy to miftake it when feparated from the plant: its tafte is fomething like that of the mulberry, but much more agreeable. It is a creeping plant, ftrikes root at each joint, and has a fmall leaf like that of the yoke-elm.

The leaf of the other plant is rather hairy, fomething like that of the mallow. The ftem which fupports the fruit is fo little elevated, that a part of it is frequently under-ground. It is fhaped like a mulberry, but of a lively vermillion: the feed is dry and almoft taftelefs.

Amongft the herbage and heath, there grows another plant full as remarkable as thofe I have mentioned. Its fruit is pleafing to the eye, and agreeable to the tafte. It makes an excellent liquor infufed only in brandy and fugar, as it has a very grateful odour of amber and mufk, which would not difguft any one who has even an averfion to thofe two perfumes, and would be infinitely pleafing to thofe who are fond of them. The Indians who inhabit the fouthern parts of Canada, prefer the infuſion of this plant to the beſt tea. They drink it both for pleaſure and health; they ſay that it cheers the heart, reſtores and fortifies the ſtomach, cleanſes the brain, and communicates a balſamic virtue to the blood. M. Duclos our captain, a Canadian, and ſome officers belonging to our frigate, who made a conſiderable ſtay in that country during the laſt war, aſſured me of this, and took great pains to provide a plentiful ſtock of it. They call this plant *Lucet muſqué.* It has the delicate and ſweet ſcent of myrtle. Its ligneous branches lie cloſe to the ground, creeping like thoſe of the wild thyme, which this plant reſembles in its ſtems and leaves, with this difference only, that they are not quite ſo acute. I never ſaw it in flower; nor do any of our officers remember to have ſeen it: but whatever its flower may be, it is ſucceeded by a fruit reſembling that of the myrtle, only larger when it comes to maturity. At firſt it appears red, and moſt commonly grows white as it ripens. It then becomes oval, and is crowned with four green points which expand themſelves like thoſe of the pomegranate. It contains a ſmall quantity of ſeeds, like the *Vitis Idæa*; its juice is ſweet. Moſt of theſe fruits are as large as that of the hawthorn, but I have ſeen ſome of the ſize of a ſloe. See the plates.

Another plant whoſe name and properties I am unacquainted with, grows in ſandy places upon the ſea-coaſt: but it is uncommon. Suſpecting that it might have ſome virtues, which, if diſcovered, might prove of advantage to mankind, I gathered the ſeed. The leaves of it, which reſemble the head of a blunted ſpear, and are nearly oval, are produced on a long ſtalk which riſes from the root itſelf. They are more woolly than thoſe of the *Verbaſcum*, called *High-taper*, or *Mullein.* Its flowers are yellow, radiated, diſpoſed in bunches, and ſupported by a calyx, which becomes round like that of an artichoke, and when the flower is fallen, contains a long angular ſeed much like that of endive.

We met with but one kind of ſhrub in that part of the country which we viſited. It is found in moiſt lands, on the

 little

little hills through which the waters paſs in their deſcent from the heights. This ſhrub grows to the ſize of roſemary, which it perfectly reſembles in its leaves, except that they are ſhorter and rather ſmaller. The flowers are white, much like thoſe of the Eaſter daiſy, or the daiſy of the fields. They are not ranged in the form of ears like thoſe of roſemary, but each flower is placed at the extremity of each ſmall branch, in ſuch a manner that the ſhrub appears intirely covered with them.

The flowers and leaves have ſcarce any ſmell; and the little they have does not reſemble that of roſemary. It is certainly not the plant which Frezier mentions in his account of the South Sea by the name of the P . . . . . an Indian name, and which the author of admiral Anſon's voyage affirms to be very common at Port St. Julian, on the coaſt of the Patagonians, ſituated in almoſt the ſame degree of latitude with the Malouine Iſlands, where the ſhrub I ſpeak of is alſo very common: but he ſays, that it reſembles roſemary and has the ſame ſmell. The bark of this on the Malouine Iſlands is greyiſh, tolerably ſmooth, and the wood is yellow.

Among the ſhrubs may be ranked a ligneous plant, which commonly grows in ſuch parts of theſe iſlands as are ſupplied with freſh running water. At the diſtance of ſome paces it might be taken for a ſmall roſe-tree; but upon a nearer examination, the leaves, which come out in pairs, rather reſemble that of pimpernel. It is indeed rather longer, and its taſte as well as the top which elevates the ſeed, bear ſome affinity to it: this top is oval, not unlike the outward coat of the cheſnut, or one of thoſe red berries which the ſweet briar or wild roſe exhibits in autumn. This ſhrub has a creeping ſtem, ſometimes an inch in thickneſs, and four or five feet in length. From this ſtem are produced branches eight or ten inches high, which are terminated by the flower and the ſeed. I did not ſee any of theſe in flower, it being too late in the ſeaſon.

The drier ſoils produce two or three ſorts of broom with a red fruit, which differ much from the European kind. They have

have all a refinous fmell. There is another pretty large plant which taftes exactly like the young fhoots of the pine-tree, called in Canada the *Sapinette*, of which is made a fermented liquor of the fame name, which is very wholefome. We tried to make the fame kind of liquor with this plant: thofe who had been in Canada affirmed that it had the fame tafte. We drank of it feveral times, and found it exceedingly good. It will be of great fervice to thofe who may hereafter fettle in thefe iflands; as this plant is to be found there in great plenty, and the liquor that is made from it may be ufed inftead of beer. The ftem and leaves are of a pale green inclining to yellow, and may be claffed with thofe creeping plants which have a round ftem very pliant, and fometimes as thick as the barrel of an eagle's quill, though oftner that of a goofe-quill The leaves come out in pairs on the fides of the branches, being faftened to a very fhort footftalk, and are fhaped pretty much like thofe of the gum-tree, which I have fpoken of before. This plant flourifhes as well in low, as in high and dry grounds. The flower, which is herbaceous, leaves behind it a white tuft fhaped like a loofe brufh, and bears no fruit.

The leaves of the largeft of the two kinds of broom which produce it, are round, and of a whitifh green; they are crowded in fuch numbers round the branches that they quite conceal them. The frnit is of the fize of a pea, of a red colour, and is tolerably well tafted.

The leaves of the other are placed in the fame manner round the branches, but are fmaller, terminating in a point, and are of a very fine green. The fruit has a kind of berry like that of the hawthorn; but its colour is a fine carmine: and it is fmaller than that of the laft mentioned broom. The plant is likewife not fo large: it is pretty commonly found among the gum-trees, between which its branches infinuate themfelves in fuch a manner that you would take it for a branch of the fame plant, though with different leaves fupporting the fruit.

This

This gum-tree forms but one green head, as its leaves do not exceed each other in length more than the fourth part of a line. It requires a very close inspection to distinguish them. They are as it were glued one above another in the form of a rose. The flower so nearly resembles the capsule which contains the seed, that it may easily be mistaken for it. This capsule greatly resembles that of the aniseed, but it is of a grey earth colour. I have seen several of these gum-trees more than ten feet at their greatest diameter, and from four to four and an half in height. They are in general nearly circular; but the largest are shaped like a potatoe cut in two.

I have met with few remarkable sea plants except that which our mariners called *Baudreux*. Its stems rise to the surface of the water, upon which they extend a great way, and are supported by means of a kind of bubble filled with air, from which the stalk of the leaf is generated.

These baudreux are found in great quantities along the coast, and even a full league from land, in places from fifteen to eighteen fathoms deep: so that the stem, in order to reach the surface and extend itself so far upon it, must be twenty fathoms in length. I once amused myself with taking measure of one which the waves had by chance broken off, and thrown upon the surface; I thought I should never have seen the end of it.

The roots of these baudreux, as well as the stem of the plant, are yellow, interwoven with each other so as to form a large bunch, which affords a shelter to the finest muscles, both of the Magellanic as well as the smooth and common kinds. Here are likewise found purpura, cochleæ, and several other shells. The mother of pearl and fasciated limas live amongst the stems and leaves.

These leaves are two feet and an half in length, and their greatest breadth is four inches. They are of a yellow red, resembling in colour the leaf of a dead tree which begins to rot. Their superficies is uneven, as if the leaf was figured. See the plates.

This

This plant produces thirty ſtems from a ſingle root, which is faſtened to the bottom of the ſea by one extremity, ſhaped like the broad end of a trumpet, or wide funnel. From this proceeds a bundle of roots or intertwiſted ſtems, among which ſtones and ſhells of the kind I have mentioned are frequently found. The leaves grow upon the ſtem at intervals. A mucilaginous and frothy fluid oozes from the ſtems, and affords nouriſhment to the ſhell-fiſh that adhere to them. When the waves have diſlodged theſe bundles from the bottom, and thrown them upon the ſhore, and when the leaves becoming withered by the action of the air, and the rays of the ſun, are ſeparated from them, our mariners call them *Goemon* or Sea-graſs. If one is not careful to take out the ſhell-fiſh as ſoon as the ſea which has thrown them on ſhore has ebbed, the ſhells are not worth preſerving: the ſun calcines them, deſtroys their fineſt colours, and reduces them to lime, ſo that they become friable between the fingers. In order therefore to collect ſuch of them as deſerve a place in the cabinets of the curious, theſe weeds muſt be pulled up from the bottom of the ſea with the drag, or the ſhell-fiſh picked out from the weeds as ſoon as the ſea has thrown them on ſhore.

The lepas, or patellæ, of the Malouine Iſlands are ſuperior in beauty to any in France. They are for the moſt part oval. The inner ſurface exhibits the fineſt mother of pearl; the bottom of the concavity is often lined with the red brown tortoiſe-ſhell, which appears to be gilt. The outward ſurface is ſtriated and channeled, the projecting parts are of a brown tortoiſe-ſhell colour, and the bottom is variegated with mother of pearl and gilt tortoiſe-ſhell.

I have ſeen ſome that were three inches and upwards at their greateſt diameter. There are five or ſix ſorts of them which are more or leſs oval; in ſome the cavity is of a conſiderable depth, in others, though of the ſame diameter, it is leſs deep by three-fourths. I have ſome in my poſſeſſion an inch and an half broad at their ſmall diameter, which are not three lines in depth:

depth; and others an inch broad whose cavity is an inch deep. The inner surface of these is most commonly of the colour of fine white porcelain, and the bottom of the cavity of gilt tortoise-shell.

There are some of this kind very large and beautiful, having an oval aperture in the center of the top, white within, and stained with stripes of purple and violet, which widen as they extend from the center to the circumference.

The fourth sort is by some called Dragoon's cap; the largest aperture I have had an opportunity of observing does not exceed from nine to ten lines in diameter, and six or seven in depth; the outward superficies is grey, almost smooth, and has sometimes stripes inclining to brown; the inside is usually of the colour of the lees of red wine with a little tincture of brown.

In many of these patellæ the perforation in their convex part is not placed directly in the middle, but rather towards one of the edges of the greatest diameter. In one of them it is situated as near as possible to one of the extremities. This patella is very flat; its shell is so thin, that it requires great nicety and care not to break it. The two surfaces are smooth, and silvered over when the external one is stripped of its outward covering, which is of a filemot colour. Here is likewise found a concamerated patella which is small and white, both within and without; I never saw any of this sort but upon the shore, and they were always without the fish. To these may be added that sort which our seamen call *Gondolas* or *Boats*; because it resembles them in figure when the bottom is turned uppermost: but their upper surface is like the coat of the millepedes. It is composed of eight pieces, inserted into each other in such a manner, that the fish can roll itself up, form itself into a round ball, and inclose itself in its shell. A fleshy substance runs quite round with rough hairs three or four lines in length. The shell is variegated with stripes or streaks of a fine bluish green, a milky-coloured white, and a darkish brown.

The Malouine iſlands abound with four kinds of muſcles; the common, the Magellanic, and two other ſorts which differ in ſhape both from the common and Magellanic. I have ſeen ſome of theſe laſt ſorts, the ſhell of which was from five to ſix inches long, and three inches broad. Thoſe which are gathered from the rocks left dry when the ſea retires, are commonly full of pearls, ſome of which are pretty enough. Thoſe which adhere to the ſhell, or are diſperſed over the body of the muſcle, are of a violet blue inclining to black; they are often uneven, and bear a great reſemblance to turnip-ſeed. The pearls of the large Magellanic kinds are white, but ſeldom of a good ſize, and clear colour. They are likewiſe very apt to break in attempting to ſeparate them from the ſhell. Thoſe which are found on the body of the muſcle, are properly nothing more than ſeeds. It is highly probable that theſe pearls proceed from ſome diſorder in the fiſh, as they are ſeldom found in muſcles which are conſtantly waſhed by the ſea-water. The want of water, at a time when the ſun darts his rays fiercely, undoubtedly occaſions an extreme thirſt, and a languor that impairs them, and creates an obſtruction; from which theſe pearls are generated.

The ſhell of one of the other two kinds of muſcles is white, tranſparent, and ſo light that the leaſt breath of air blows it off the hand. The other though larger, is of a very ſhining red brown gold colour, particularly when under water, and the ſun ſhines upon it. When empty, it is ſcarce heavier than the preceding one, for the wind alone throws it upon the ſhore. See the plate, &c.

The large and ſmall Magellanic muſcles are of the whiteneſs of mother of pearl, divided by purple ſtripes, adapted to the circular figure of the ſhell. The coat which covers the external ſurface is of a muddy brown: but when this is taken off, it diſplays a fine ſky-blue veined with purple ſtripes. The channels diminiſh inſenſibly as they approach the ſharp end, which is fine mother of pearl, and from which they procced as from their center. See the figure of theſe muſcles in the plate.

A great number of other different ſhells engage the attention of the curious upon the coaſt of theſe iſlands: foliated buccina, ſpinoſe buccina, ſcrew ſhells, of different kinds, Purpura, faſciated Cochleæ, concamerated Cochleæ, Neritæ, ſmooth Chamæ, ſtriated Chamæ, Scollop ſhells, Pectines, Echini, Sea-Aſteriæ, and a ſpecies of Concha, which our ſeamen call *Gueulle de Rayes*. This laſt ſhell has not till lately been known except among the foſſil ſhells, and it has been doubted whether it exiſted in nature. In the ſubſequent voyages made to the ſame iſlands, ſo great a quantity of them has been collected, that they have been diſtributed among the cabinets in Paris: ſo that the only ſhell of the kind which I depoſited, on my return, in the cabinet of the Abbey of St. Germain des Prés, is no longer a rarity.

There are probably ſeveral other ſhells along the coaſt of the main ſea, which I have not had an opportunity of ſeeing, becauſe the place where we anchored was about ſix leagues in the bottom of the bay; and the ſpot on which we pitched our tents, and fixed our habitation was near two leagues farther. Throughout this whole bay, I have ſeen no other kinds of ſhells than thoſe I have deſcribed: nor did we meet with any fiſh beſides thoſe I have mentioned, except ſome white porpoiſes, and ſeveral whales.

There are three kinds of amphibious animals very commonly found on theſe iſlands; ſea-wolves, ſea-lions, and penguins. I have ſaid ſomething of each of theſe; but ſhould add, with regard to the ſecon hat the name of *ſea-lion* does not ſo properly belong to thoſe I have deſcribed, (and of which the author of Admiral Anſon's Voyage treats pretty largely) as to another ſpecies, in which the hair that covers the back part of the head, neck and ſhoulders, is at leaſt as long as the hair of a goat. It gives this amphibious animal an air of reſemblance to the common lion of the foreſt, excepting the difference of ſize. The ſea-lions of the kind I ſpeak of, are twenty-five feet in length, and from nineteen to twenty in their greateſt circumference. See the plate. In other

other reſpects they reſemble the ſea-lions, of which I have given the figure. Thoſe of the ſmall kind have a head reſembling a maſtiff's with cloſe cropt ears.

The teeth of the ſea-lions which have manes, are much larger and more ſolid than thoſe of the reſt. In theſe all the teeth which are inſerted into the jaw-bone are hollow. They have only four large ones, two in the lower and two in the upper jaw. The reſt are not even ſo large as thoſe of a horſe. I brought home one belonging to the true ſea-lion, which is at leaſt three inches in diameter, and ſeven in length, though not one of the largeſt. We counted twenty two of the ſame ſort in the jaw-bone of one of theſe lions where five or ſix were wanting. They were intirely ſolid, and projected ſcarce more than an inch, or an inch and an half beyond their ſockets. They are nearly equal in ſolidity to flint, and are of a dazzling white. Several of our ſea-men took them for white flints when they found them upon the ſhore. I could not even perſuade them that they were not real flints, except by rubbing them againſt each other, or breaking ſome pieces off, to make them ſenſible that they exhaled the ſame ſmell as bones and ivory do when they are rubbed or ſcraped.

Theſe ſea-lions that have manes, are not more miſchievous or formidable than the others. They are equally unwieldy and heavy in their motions; and are rather diſpoſed to avoid than to fall upon thoſe who attack them. Both kinds live upon fiſh, and water-fowl, which they catch by ſurprize, and upon graſs. They bring forth and ſuckle their young ones among the corn-flags, where they retire at night, and continue to give them ſuck till they are large enough to go to ſea. In the evening you ſee them aſſembling in herds upon the ſhore, and calling their dams in cries ſo much like lambs, calves and goats, that, unleſs appriſed of it, you would eaſily be deceived. The tongue of theſe animals is very good eating: we preferred it to that of an ox or calf. For a trial we cut off the tip of the tongue hanging out of the mouth of one of theſe lions which was juſt killed. About ſixteen or eighteen of us eat each a pretty large piece, and we all thought it ſo good, that we regretted we could not cut more of it.

'Tis ſaid that their fleſh is not abſolutely diſagreeable. I have not taſted it: but the oil which is extracted from their greaſe is of great uſe. This oil is extracted two ways; either by cutting the fat in pieces and melting it in large cauldrons upon the fire; or by cutting it in the ſame manner upon hurdles, or pieces of board, and expoſing them to the ſun, or only to the air: this greaſe diſſolves of itſelf, and runs into veſſels placed underneath to receive it. Some of our ſeamen pretended that this laſt ſort of oil, when it is freſh, is very good for kitchen uſes: this, as well as the other, is commonly uſed for dreſſing leather, for veſſels, and for lamps. It is preferred to that of the whale: it is always clear, and leaves no ſediment.

The ſkins of the ſea-lions are uſed chiefly in making portmanteaus, and in covering trunks. When they are tanned, they have a grain almoſt like Morocco. They are not ſo fine, but are leſs liable to tear, and keep freſh a longer time. They make good ſhoes and boots, which, when well ſeaſoned, are waterproof.

The Penguin is ſo ſingular an animal, that it is not eaſy to ſay to what genus or ſpecies it belongs. It has a bill like a bird, and feathers; but they are ſo fine and ſo unlike common feathers, that they have properly the appearance of hair as fine as ſilk, even when you are near enough to examine and touch them. You can only be convinced of the contrary by plucking one of them, upon which you diſcover the barrel and feathers of a quill. Inſtead of wings it has two fins, which are articulated in the ſame manner as the wings of birds, and are covered with very ſmall feathers which might be taken for ſcales. At firſt ſight it appears to have no thighs, and its feet, which are rough like thoſe of geeſe, ſeem to come out directly from the body on each ſide of the tail, which is nothing more than a continuation of the feathers, nearly in the ſame manner as in ducks, but much ſhorter. The neck, the back, and the fins are of a bluiſh grey, blended through-

throughout with a pearl-coloured grey. The belly down from the neck is white. The old ones have a white ſtripe round their eyes mixed with yellow, which is not unlike ſpectacles. From thence this ſtripe extends on both ſides along the neck, where it is ſometimes double, and paſſing cloſe to the fins, terminates at the feet which are of a darkiſh grey, and have very thick toes. Its noiſe is like the braying of an aſs. Its aſpect and its motion are different from that of birds. It walks upright, with its head and body erect, like a man. At the diſtance of an hundred paces, you would take it for one of the children of the choir in his habit. The largeſt of thoſe we have taken may be about two feet ten inches high.

They live among the corn-flags like the ſea-wolves, and earth themſelves in holes like foxes. They ſuffer one to come ſo near them without ſtirring, that one may kill them with a ſtick. As you approach them, they look at you, turning their head to the right and then to the left, as if they made a jeſt of you, and muttered ironically *What a fine fellow have we got here!* They ſometimes retreat when you are five or ſix feet from them, and run pretty much like a gooſe. If they are ſurpriſed and attacked, they run in upon you, and endeavour to defend themſelves by ſtriking at your legs with their bills; they have recourſe to ſtratagem to gain their point, and pretending to retreat ſideways, turn back in an inſtant, and bite ſo hard that they take the piece out, if you have nothing to ſecure your legs. They are uſually ſeen in flocks, ſometimes to the number of forty, ranged in order of battle, and eye you as you paſs at the diſtance of twenty paces. Their fleſh is black, and has rather a perfumed taſte. We ate of them ſeveral times in ragouts, which we found to be as good as thoſe made of a hare. We took off the ſkins from ſeveral, with a view of preſerving them, but they were ſo oily that we threw them into the ſea: it was likewiſe their moulting-ſeaſon. I wrapped the ſkin of a young one in ſtraw, which is in very good preſervation: I have depoſited it in the cabinet of natural

ral curiofities belonging to the Abbey of St. Germain des Prés. See the Plate.

When they take to the water, and find it deep enough to cover their neck and fhoulders, they plunge into it, and fwim as quick as any fifh. If they meet with any obftacle, they fpring four or five feet out of the water, and then plunge again, in order to purfue their courfe. Their dung exhibits only an exceeding fine earth, of a yellowifh red, interfperfed with fmall fhining points like mica; it might be taken for the *Lapis fortuitus.*

As for the birds in thefe iflands, there are but few found upon land. There is a flock of birds upon the fhore like fmall thrufhes, of a brown grey, fo tame that they come flying almoft upon your finger. I killed ten with a fmall fwitch in lefs than half an hour, without changing my place. They fcratch among the fea-grafs which is thrown on fhore, and eat the worms and fmall fhrimps, which we call *Púces de mer*, becaufe they are fkipping inceffantly like fleas.

Blackbirds are likewife found here, and a kind of thrufh with a yellowifh belly. They feed in the fame manner as the bird I have juft now defcribed. We killed a kind of ftarling in the fields, the upper part of whofe neck, back, and wings, is marked and fpeckled almoft like thofe in France; the bill is likewife fhaped in the fame manner: but the lower part of the neck and belly are of a very fine red, fomewhat inclining to a flame colour; this red is befprinkled with fome black fpots. I could not give a true refemblance of it without making ufe of the minium or red lead. See the Plates.

Here are great numbers of wrens like thofe in France, fnipes, curlieus and fea-larks: likewife a fmall bird, not often feen, like thofe that hover about flocks of fheep; thefe birds have all an excellent flavour.

Upon the fea-coaft is almoft always feen a kind of duck, which flies in pairs and fometimes in flocks: the feathers of its wings are very fhort, and only ferve to fupport it in running upon the water, for it never flies. Its plumage is grey, its bill and feet

5 yellow.

yellow. When it is not ſhot dead, it continues its flight upon the ſurface as long as the leaſt breath of life remains. Its fleſh is oily, and has a fenny taſte: it was eaten however by our ſhips companies when no buſtards were given them. Theſe ducks uſually weigh at leaſt between nineteen and twenty pounds each. We called them grey geeſe, to diſtinguiſh them from the kind which affords that fine down of which muffs are made. They are not better eating than the ducks; their fleſh has even a diſagreeable ſmell, which their oily ſkin retains a conſiderable time, though expoſed to the air. This diſguſting circumſtance prevented our making a collection of them. They may probably be of the kind called *Cahuitahu du Para*.

The large feathers in their wings are of an iron grey, the ſmall ones a mixture of green and gold, and vary their colour like thoſe of a wild duck; the reſt of their body is white. The articulation of the wing is armed with a ſpur as hard as horn, not very ſharp, but rounded like a cone, and about half an inch long. Their bill and feet are black. The ſtrokes they give with their wings in defending themſelves, are accompanied with ſuch force that they bruiſe the fleſh where the blow lights. The buſtards are likewiſe armed with a ſpur of the ſame kind. I received a blow upon my hand from one that was even mortally wounded with ſhot. I felt a very acute pain for a full quarter of an hour, and the mark of the bruiſe remained more than two days.

The wild Ducks, which are here very common, are like thoſe of France, but not near ſo good; having, in general, the taſte of muſcles: but the Teal and the Divers, which are no leſs numerous, are excellent.

There are likewiſe found prodigious numbers of another ſpecies of Divers which are tolerably good, though they have rather an oily taſte. Our ſeamen called them at firſt *Becſics*, and afterwards *Coyons* and *Nigauts*, becauſe they ſuffered themſelves to be killed with ſtones, not attempting to fly away unleſs they were hit. They aſſemble in flocks upon the rocks near the ſea-coaſt, ſometimes to the number of an hundred and upwards. When

we

we went on ſhore in the ſloop, ſeveral companies of them, conſiſting of two or three hundred, paſſed only eight or ten feet above our heads. There are three kinds of them; all nearly of the ſame ſize. Some are quite black; in others the fore-part of the neck and all the belly is white: in the third kind, the belly and breaſt is white, and the reſt black. Their bill, which is of the ſame length as their head, is black and ſharp, like that of birds which are not aquatic. Their feet are of a dark grey and webbed; but inſtead of four toes, they are only furniſhed with three, which differ in ſhape from thoſe of other water-fowl. See the Plate. Our ſeamen preferred them to wild ducks; and indeed they had not near ſo diſagreeable a taſte.

Red-ſhanks and ſea-pies are very good here; but the buſtards in particular are exquiſite, either boiled, roaſted, or fricaſſeed. It appeared from the account we kept that we ate fifteen hundred. It is indeed hardly to be conceived, that the ſhip's company of our two frigates, conſiſting of an hundred and fifty men, all in perfect health, and with good ſtomachs, ſhould have found a quantity of theſe birds ſufficient for their ſubſiſtence during a ſtay of more than two months, within a tract of country not ceeding three leagues.

Theſe are almoſt all the kinds of animals we ſaw in that part of the iſland where we fixed our encampment, except two or three kinds of ſmall birds, ſome of which reſemble the Siſkin, others the linnet; and a kind of wagtail that has not ſo long a tail nor ſuch darkiſh ſtripes as thoſe of France. There is likewiſe found a kind of white gull, and a carnivorous bird of the ſize of a common hen, with reddiſh grey plumage. The people on board called them grey gulls. They came very near us, and, when we were in purſuit of game, flew ſo cloſe to our heads that more than once they ſwept off the caps and hats of our people. They make a noiſe much like a duck; and though they are water-fowl their feet are not webbed; but they ſeize their prey voraciouſly, by means of the very ſharp talons with which their toes are armed; and when their prey is either not large enough

enough to ſupport upon the ſurface, or too heavy for them to carry off, they tear it in pieces with their beak and talons, flapping their wings all the time. They ſettle however upon the water, and remain upon it like ducks, but I never ſaw any of them dive. Nobody thought it worth while to ſhoot them, concluding they would be very bad eating.

A ſmall heron with a crown is likewiſe found here, whoſe feathers are of an aſh-coloured bluiſh grey: the crown is compoſed of three white feathers three inches long, reſembling in ſhape the crown of the peacock. Upon the breaſt, round the neck, under the wings, on the lower part of the back, and under the thighs, there is a down, part white and part of a citron coloured yellow, at leaſt an inch long, exactly reſembling a piece of the fineſt raw ſilk.

In the ſecond voyage we ſaw ſome paroquetes, and a kind of ſwan with a red bill, the whole neck being of a moſt beautiful black, and the reſt of the plumage white.

The climate and the air appear to be ſo wholeſome, that all the perſons we left behind remained on the iſland of their own accord, and live under the ſame roof, in the apartments provided for them in the building erected near Fort St. Louis; where they will ſubſiſt as well by ſhooting, as on the proviſions with which their magazine is plentifully ſtored for two years. One of the two Acadian families which we carried over conſiſts of the huſband, his wife, two children, one a boy three years and an half old, the other a girl about a year old, and of two young women, ſiſters of the mother, the eldeſt nineteen, and the youngeſt eighteen. The other family conſiſts of the huſband, his wife, who is pregnant and ready to lie in, a ſon four years old, and the mother's ſiſter, a girl of ſixteen.

There remain here all kinds of artificers, as ſmiths, iron-mongers, carpenters, joiners, maſons, bricklayers, ſhoemakers, bakers, a captain of a ſhip, ſailors, &c. The ſoil is very promiſing, and it is very probable that theſe artificers will make good uſe of the tools and ſeeds of all kinds that were left them, and that this colony will flouriſh, if the miniſtry make a point of improving it. Beſides proviſions, we left ſeven heifers and two young bulls, eight hogs and two boars, a few ſheep, a goat, two horſes and a mare, which range about the country.

We ſaw no kinds of reptiles or noxious inſects here, only ſome ſmall common flies, ſome ſmall field ſpiders called *Spinners*. There is no kind of quadruped except the ſmall wolf or fox I have mentioned. This will not be ſurpriſing if we recollect that travellers aſſure us, that no reptiles or inſects are found in the ſouthern part of Chili, which is pretty nearly in the ſame latitude with, and almoſt oppoſite to the Malouine Iſlands. See the chart of the coaſt which we made a ſurvey of from our place of landing on the three iſlands, which we took at firſt to be the Sebalds, to the port or eaſtern bay where we anchored.

In the ſecond and third voyages the ſhips, in returning from the ſtreights of Magellan, ſailed along the ſouthern coaſt of theſe iſlands, as it will be deſcribed in the extract of the journals of M. Alexander Guyot, and M. de Bougainville; and they have made a chart of it, which is given in the plates.

On Tueſday the 10th of April, we perceived ſeveral whales and a great number of birds, among which were ſome Petterils, called *Damiers*, or cheſs-boards, on account of their plumage being chequered with black and white. The head and part of the neck, likewiſe the tip and middle of the wings, are black; the reſt of the body is not white, though it appears to be ſo at the diſtance of piſtol-ſhot. On a nearer view you find that the extremity of the wings is black; they have the appearance of round ſcales edged with black. It is of the ſize of a large pigeon. As we had now very freſh gales, the rolling was ſo conſtant and violent, that it was impoſſible to keep the diſhes upon the table without holding them, and every perſon was obliged to have his plate in one hand and his fork in the other.

Notwith-

Notwithſtanding every poſſible precaution, a ſoup-diſh, ſome plates and drinking glaſſes were broken at dinner. Theſe rollings were ſo violent in the night time, that thoſe who did not lie in cots or hammocks, could not reſt in their beds.

On the 11th, the ſame weather continued all the morning. In the afternoon we ſaw ſeveral birds and whales. We were ſtill ſo exceſſively rocked with the rollings, that it was almoſt impoſſible to keep the deck. We ſaw a quantity of birds, and a very large whale, which accompanied the ſhip for a conſiderable time, at the diſtance of a muſket-ſhot.

On the 14th, we ſaw a number of grey ſea-mews, and ſome Quebrante-Ueſſos.

From the 14th to the 24th, nothing remarkable happened: but on this day, ſoon after ſeven in the morning, we ſaw a flying fiſh by ſome called *Adonis*, but for what reaſon I cannot ſay. There are ſeveral kinds of them. Some are diſtinguiſhed by the colour, others by the length of their fins which ſerve them as wings. There is a third kind with four wings inſtead of two, which is the uſual number. None of thoſe we caught between the tropics had more than two wings, ſome of a larger, others of a ſmaller ſize. They were all of a fine deep blue, ſilvered over on the back to half the breadth of their body, and the whole belly was of a very bright blue, ſilvered over in the ſame manner. The largeſt of them which fell into our frigate was eight inches in length, including the head and tail. In ſome the wings were only two inches long, in others they extended as far as the tail.

Few animals have ſo many enemies as the flying fiſh. They ſpring out of the ſea to eſcape being devoured by thunnies, bonitos, ſharks, &c. and in the air meet with birds that are always upon the watch for them. They riſe ſo high above the water that they ſtrike againſt the ſails and ſhrouds of ſhips into which they fall: and this is the only way of catching them. Their fleſh is good and delicate. You ſee them ſpring out of the

the water by hundreds like flocks of larks, at which time their wings make them appear white.

On the 25th in the morning, we paffed the Tropic of Capricorn, and entered a calm and warm climate. Accordingly laft Sunday, which was Eafter-day, we all put on our lighter clothing.

On the 27th, we fpied land before us, and fteered N. E. $\frac{1}{4}$ N. then directed our courfe fo as to pafs within half a league of it. At half an hour paft fix we found by obfervation that this land was the ifland of Afcenfion, which appeared to us as defcribed in the plate. We computed its diftance at about fix leagues. As we approached this ifland, it appeared to me to be compofed of feveral rocks joined together, or of a fingle rock having different fummits, between which there was a little earth or fand, covered here and there with fome herbage, which gives a little verdure to the floping declivity which runs down to the fea on the fide towards the N. E. and E. N. E. At half an hour paft nine, feveral of our people fancied they faw trees, but on taking an accurate view of the whole with perfpective glaffes; we concluded that what had the appearance of trees was nothing more than broom or fhrubs. At the bottom of the declivity juft now mentioned, you fee a kind of fandy flat, flightly covered with verdure, upon the coaft and to the N. N. E. of the ifland. At half an hour paft eight, we were at two leagues diftance or thereabouts.

At eight o'clock, we difcovered another ifland to the E. $\frac{1}{4}$ N. E. about fix leagues from the ifland of Afcenfion; three fmall iflands foon after appeared clofe to the laft difcovered one.

This ifland, with the iflets about it, might probably be the fame that fome feamen paffing too far eaftward, and not feeing the ifland of Afcenfion, have called Trinity ifland; fince many navigators pretend that the ifland of Afcenfion, and Trinity ifland are one and the fame; at leaft an ifland, and three rocks or iflets, are found E. $\frac{1}{4}$ N. E. of that of Afcenfien, as they are laid down in the charts. This ifland of the

Trinity will then be the largeſt, or rather the largeſt of the four iſlets I have been deſcribing. Indeed the latitude in which Trinity iſland is placed, and the latitude of Aſcenſion, would not be found to coincide: but the charts, which are ſo faulty in placing other iſlands, may poſſibly be miſtaken in the poſition of theſe. The largeſt of the iſlets, which I took for Trinity iſland, was ſeen by us at the diſtance of five leagues or thereabouts, but appeared leſs extenſive than the iſland of Aſcenſion, when obſerved at the ſame diſtance. The two iſlets, or rocks, at firſt view had greatly the appearance of ſhips under ſail.

We paſſed between theſe two iſlands without altering our courſe; and ſaw nothing but ſteep rocks, ſeveral of which appeared almoſt perpendicular. We ſaw no inhabitants but ſea-fowl. As we coaſted ſo near the land, we ſhould have ſeen ſome tortoiſes if there had been any in theſe iſlands. One of theſe animals, a foot or rather leſs at its longeſt diameter, paſſed alongſide our frigate four days before: but we were then at too great a diſtance from theſe iſlands to think it probable that it came from thence.

On Sunday the 29th, in the morning, we ſaw ſome flying-fiſh, ſome birds called *Taylor Birds*, others called *Frigates*, and ſome Tropic birds, called *Paille-en-Cul*, or otherwiſe *Flêche-en-Cul* and *Fêtu-en-Cul*. The ſailors, who name things according to their ideas, give the bird this name on account of the two feathers of its tail which are very long. Some of this kind, which hovered over our ſhip for a conſiderable time, appeared to be of the ſize of a pretty large red partridge. The Tropic bird has a ſmall well-ſhaped head; its bill is about three inches long, pretty thick and ſtrong, rather crooked, but pointed and red like its feet, which are webbed. The wings are very large in proportion to the body; and indeed this bird flies very well, and to a great height. It makes excurſions three or four hundred leagues from land, reſts upon the water, and lives upon fiſh.

Its

Its plumage appears to be intirely white. Our ſeamen who have had a near view of it aſſure me, that it is variegated with white and blue. The tail conſiſts of twelve or fifteen feathers from five to ſix inches long. The two middle ones, which are from fifteen to eighteen, are joined in ſuch a manner that they appear to be one.

Some of our officers who had been at the iſland of Mauritius, or the iſle of France, communicated to me a ſingular obſervation they made there, that the Tropic birds never appeared in the port of that iſland, except on the very day, or about twelve hours before the arrival of ſome French veſſel. Accordingly, when one of theſe birds is ſeen, the inhabitants are in a manner certain that a ſhip will come into port ſoon after.

On the 4th, 5th, and 6th, we ſaw many flying fiſh; and on the 8th, a great number of porpoiſes paſſed very near our veſſel; we endeavoured to harpoon them, but without ſucceſs.

On Thurſday the 10th, in the evening, we caught a ſhark, ſaw a quantity of porpoiſes, ſome thunnies, and ſeveral bonitos.

On Friday the 11th, at ſix in the morning, we caught a ſhark. We attempted to catch ſome thunnies with the loſs of two hooks, which were thicker than the barrel of a gooſe-quill. One of theſe thunnies broke two of them which were faſtened to the ſame line. The bonitos would not bite.

On the 12th, we caught one porpoiſe among a prodigious number, and a ſhark; at three o'clock we caught a bonito, in the belly of which was found a fiſh called *Cornet*, which it had probably juſt then ſwallowed, as it was ſtill intire, and preſerved its natural colours. I immediately made a drawing of it, repreſented in the plate.

The reader muſt not form his idea of the ſize of this fiſh from the figure I have given of it. In the opinion of the ſeamen who frequent the South Sea, the *Cornet* is the largeſt of all ſea-fiſh. It ſeizes its prey by the aſſiſtance of the moveable claws at the end of its ſnout. Theſe ſeamen likewiſe tell us, that faſtening upon, and catching hold of ſhips with theſe

5 claws,

claws, it climbs along the tackling: that if it does this in the night-time unperceived, its enormous weight throws the ſhip ſo much upon her ſide that ſhe is in danger of being overſet. They accordingly take great care to keep a good look out, with hatchets, and other ſharp inſtruments to cut the claws of this fiſh, as ſoon as they ſee them faſtened upon the ſhip Our captain, and his brother Alexander Guyot, who have made ſeveral voyages in the South Sea, confirm this account; but add, that they never ſaw any of ſuch an immoderate ſize, that they have taſted ſome that weighed an hundred and fifty pounds, and had an excellent flavour. If one may judge of them by the ſmall one deſcribed in the plate, they muſt be very delicate. The ſcales of this fiſh, which are a kind of ſheath to it, as well as the fiſh itſelf, were almoſt tranſparent.

On Sunday the 13th, we continued to ſee a great number of porpoiſes all the morning, and a large ſhark which would not take the bait.

On the morning of the 14th, we ſaw ſeveral bonitos, thunnies, and a great number of flying fiſh, ſeveral of which falling into the ſhip made an excellent diſh for dinner.

On the 15th, about four o'clock, we caught two thunnies, and two bonitos with the harpoon.

The thunny is a fiſh well known in the Mediterranean. But whether the deſcription which M. Valmont de Bomare gives of it, in his dictionary of natural hiſtory written on the plan of Lemery's, is not exact, or whether the fiſh whoſe figure I give in the plate is not the thunny, or whether the thunny caught between the tropics differs from that of the Mediterranean, it does not agree with M. Valmont's deſcription. In thoſe we caught, the ſcales are neither large nor broad, nor the back of a blackiſh caſt, but of a fine deep blue, which brightens inſenſibly towards the fins, which are ſhaped like ſcythes, and placed near the gills. Theſe two fins, as well as the two ſmaller at the bottom of the belly, are of a very deep grey, or bluiſh black inclining to grey. That on the back, and the two on the belly,

ſituated

ſituated at about two-thirds of the length of the body, are of a gold colour, as well as ſome parts reſembling the teeth of a ſaw and extending from the fins to the tail, which is arched. On the outſide, their gills do not appear to be double. Their ſnout, which is not thick, is pointed, and has ſmall teeth that are very ſharp. M. Valmont ſays, that this fiſh dies ſoon after it is out of the water. That which I delineated, lived near half an hour hung up by the tail near the main-maſt. It would probably have lived much longer, if, by ſtruggling to diſengage itſelf, it had not diſgorged its heart, which fell upon deck in my preſence, and on taking it into my hand continued its palpitation near a quarter of an hour. In voiding it, it diſcharged a great quantity of blood through its gullet, ſome drops of which I have repreſented upon the ſurface of the lower jaw. Its fleſh is ſomething like veal; but drier and more firm.

The bonito is a large fiſh: its figure from the head to three-fourths of its length is round; from thence it begins to grow flatter, and terminates in a pretty thick tail, which is forked, and like that of other fiſhes. As it has little or no neck, it has two fins which are pretty long, but not broad in proportion to the bulk of the bonito. On the back is a fin, which, as it approaches towards the tail, ſeems to form there, as well as on the oppoſite part under the belly, triangular projections of a gold colour. On each ſide are placed two other fins of a blue colour, terminating in a point at the tail. Two ſmall fins appear under the belly. The back is of a very deep blue, which grows brighter towards the middle of the body. The belly is white with a caſt of greeniſh yellow, and is variegated with ſeveral ſtripes of a greyiſh colour, which ſeem to be blended together promiſcuouſly. The eye is large, and has a circle of gold round the pupil. Its head is not ſo long as that of the thunny. It is neceſſary to lard it well, as its fleſh is very dry. See the plate.

Theſe fiſh always appear in ſhoals; the ſea ſometimes ſeems to be intirely covered with them. They are caught with the ſpear, or a hook baited with an artificial flying-fiſh. The fleſh of of the bonito, which is taken on the coaſts of the kingdom of Angola, is ſaid to be hurtful. In the middle of the fleſh of ſome we found ſome live worms. They were white, of the thickneſs of the barrel of the feathers in a hen's wing, and about four lines in length.

On the evening of the 25th, we again met with ſome ſea-graſs, which the ſeamen call *Goemon à grappes de raiſin.* I have already obſerved, that the ſeeds with which it abounds are ſmall bladders, of the ſize of the largeſt ſwan-ſhot. They are not collected into ſeparate cluſters, but diſperſed over the ſtems and branches. When the ſeeds grow dry, they dwindle to the ſize of a middling pin's head. The leaves which are very ſmall, almoſt like thoſe of parſley piert, become brittle. Some of the ſtems, and a great number of the ſeeds, are incruſted with a very ſmall kind of ſhell, or ſpawn of fiſh, which is white and hard, and when rubbed againſt wood acts as a file, or the herb called ſhave-graſs.

On the 26th, in the morning, we ſaw ſuch a prodigious quantity of the ſea-graſs I have been ſpeaking of, that the ſea was almoſt covered with it. Among ſome large bundles of it, we found crabs of different ſizes, of a bright red, marked with brown ſpots. They have eight feet, and two claws. The body, or cuiraſs, is almoſt ſquare on the ſide of the head. The eyes project from the extremity of the two angles that form this ſquare. See the plate.

Several beds of this ſea-graſs, ſome of which were almoſt as broad as our veſſel, and longer, paſſed cloſe to our frigate. They are ſaid to come from the coaſts of the Canary Iſlands; others pretend that they are diſlodged from the bottom of the ſea. This opinion ſeems to be the moſt probable, ſince all the Canary Iſlands could hardly produce the prodigious quantity which appeared for fourteen or fifteen days paſt.

On the 31ſt, in the morning, being Aſcenſion-day, the weather being calm, after ſaying maſs we ſeized this opportunity of ſcraping and breaming the veſſel. At four in the afternoon we

 ſaw

ſaw a ſail which ſeemed to be ſteering W. N. W. at the diſtance of about ſix leagues. We loſt ſight of it at night.

On Sunday, the 3d of June, the ſea-graſs, which had not been ſeen for a day or two, appeared again in large quantities, and a whale of middle ſize played round the ſhip a quarter of an hour, at the diſtance of gun-ſhot.

A few days after we ſaw a bird, which our ſeamen call *Equeret*, and another called the tailor-bird hovered about our veſſel.

On the 13th, we caught a fiſh with our drag-net called *Grande Oreille*. It reſembles the bonito in every particular except the two fins, which are ſituated near the gills. Theſe fins are falciforn, and are at leaſt as large as thoſe of the thunny Its fleſh is not ſo dry.

On the 15th, at five in the morning, we diſcovered a ſail bearing N. W. of us, which appeared to ſteer the ſame courſe. Upon this we clued up our ſails, hoiſted our flag and broad-pendant, and made a ſignal by firing a gun. After ſailing as near the wind as poſſible, in order to wait for her, ſhe likewiſe hauled the wind, and continued at two cannon-ſhot or thereabouts to windward. Concluding that they did not hear the firſt gun, we fired another to windward; when they diſplayed a flag of peace, and fired a gun. Perceiving afterwards that they took no further notice, we hoiſted the enſign with a waft; which ſhe equally diſregarded, and always kept to windward nearly at the ſame diſtance. As ſhe ſailed at leaſt as well as our ſhip, ſhe doubtleſs depended upon her ſailing. We reſolved to give over the chace, as it would have diverted us from our courſe.

The French Captain could not make any proper excuſe for diſobeying the King's regulations reſpecting the marine, by which every ſhip belonging to the nation is obliged to bring to, when a King's ſhip gives the ſignal by firing a gun, and hoiſting the broad-pendant on the proper maſt, according to the rank of the commander. We went ſtill farther, by hoiſting the enſign with a waft, which is a ſignal of diſtreſs agreed upon by all civilized nations.

This

This proceeding therefore, for this reaſon at leaſt, deſerves the higheſt cenſure; and ſets a very bad precedent. Had we unfortunately been in actual danger, he would have ſuffered us to periſh before his eyes without giving us that aſſiſtance which the laws of humanity require in ſuch circumſtances.

The royal navy has ever been jealous of the trading branch. The former entertains prejudices which ſet it above the ſeamen's employment, and does not think practice neceſſary to attain a knowledge of it. The latter, inured to the hardſhips and fatigues of ſea, juſtly concludes that to excel in the nautic art, requires the practice of a whole life. Hence that party-ſpirit of which the ſtate becomes the firſt victim, ſince the liberty of the Reds, or officers of the royal navy, occaſions the ſervitude of the Blues. If we look into the annals of the laſt war, we ſhall find French privateers whoſe courage and intrepidity ſeemed to rouſe the winds to fight on their ſide. Could it be believed that a Captain of a veſſel (M. de L.) was a calm ſpectator of an engagement between a French ſnow and an Engliſh privateer, and contented himſelf with commending the conduct of the Frenchman, who, attentive to every particular, exerted every effort of ſkill and perſonal bravery, to prevent the enemy from boarding him. In ſhort he ſaw her diſmaſted, and obliged to ſtrike, without firing a gun in her defence. How eaſy was it for the Commander of a ſhip of the line well armed to ſave the brave Captain of the merchantman, and to make himſelf maſter of the Engliſh privateer! It is plain then that he remained inactive only becauſe it is not the mode to waſte any powder to promote the commerce of the nation, or protect a privateer belonging to a good citizen.

It may be urged, in excuſe for the conduct of the Captain of the French trading veſſel, in not bringing to even when we hoiſted our flag of diſtreſs, that having probably no more proviſions and rigging than were ſufficient for his own uſe, he apprehended that, if we ſhould happen to be in want of them, we ſhould ſeize his ſtores by force, if he refuſed to grant us a voluntary ſupply. This is an abuſe too common in the royal navy, and has given much offence to the trading branch: which, finding itſelf deſpiſed and ill-treated, is glad of every opportunity to retaliate; and I dare ſay would take ſome kind of pleaſure in their deſtruction, in hopes of being releaſed by it from the tyranny they experience from the royal navy. It would be for the intereſt of the ſtate if matters were ſo preciſely regulated by the royal authority, that no perſon of either party ſhould tranſgreſs the order on any pretence whatſoever, but ſhould be puniſhed with the utmoſt ſeverity. While this animoſity, perpetuated by the contempt which the royal navy expreſſes for the trading branch, and by the abuſe of its power, continues to ſubſiſt between theſe two bodies of men, the ſtate muſt unavoidably be expoſed to very great inconveniencies.

We are not diſpoſed to follow the example of others, nor is any one inclined to follow ours. This ſpirit of ſingularity, which runs through all our conduct, always tends to our deſtruction. We imitate the Romans, who employed only their freedmen in maritime affairs, and confined the land ſervice to the patricians. The Engliſh have better notions: among them the profeſſion of a ſailor is in great eſteem; and is the nobleſt of all the arts, becauſe it is exerciſed by the principal nobility in the kingdom. The French indeed differ from the Romans, not for the ſake of imitating the Engliſh, but in order to gratify a number of intereſted individuals, whoſe opinion very improperly paſſes for that of the nation. In France the art of navigation is eſteemed a vulgar employment, though the command of ſhips of the line is a poſt of honour which can only be filled by a perſon who is actually a man of ſome family. Accordingly there is more parade than ſcience or connection in our navy; there is more ſhew than real ſkill in our officers, and at the ſame time their emoluments are not adequate to their expences. In London, the idea of merit and reward is not regulated by faſhion and court-influence, but by the good of the ſtate. Nothing is regarded but merit in a ſailor whatever his condition may be. If he is a good ſeaman he is every thing, he is conſidered as a uſeful man, is employed

in

in honourable ſtations, and is rewarded in proportion to his ſervices. It were much to be wiſhed that we would think, or rather act, in the ſame manner. Virtue and merit ought to be the foundation of true nobility.

This is the ſource of that ſpirit of party which is a diſgrace to the Reds, an injury to the Blues, and the misfortune of the French government.

The reader, I am perſuaded, will excuſe this digreſſion, as it is dictated ſolely by my zeal for the public good, and the love I bear my country. It is certain, notwithſtanding the rivalſhip ſubſiſting between the two nations, that an Engliſh Commander would have been ſo far from acting like the French Captain, that he would have made what haſte he could to join us, and to give us all the aſſiſtance in his power: which conduct ever ought to be obſerved between all nations without diſtinction. We went up to and offered our aſſiſtance to the Captain of a Dutch veſſel, we had met in October laſt, which had loſt her maſts.

On the 16th we ſaw ſeveral whales, and a kind of thornback, which our ſeamen call *Rouet*. We kept ſight of the French ſhip before mentioned both yeſterday and to-day: it continued the ſame courſe from the E. $\frac{1}{4}$ N. E. at the diſtance of about three leagues from us, and we had proceeded about as far in our courſe.

On the 18th, at half paſt five in the morning, we ſpied a ſail coming from the Eaſtward. At eight it came abreaſt of us, and we ſpoke with it. It was the St. Paul de Grandville, Captain Deſveau, bound to Newfoundland.

On the 20th and 22d we ſaw ſeveral ſhips; and on Sunday evening the 24th we ſteered S. E. $\frac{1}{4}$ E. with a gentle breeze from N. N. W. to W. N. W. in order to obſerve the land, which we ſaw at ſix in the evening.

On Monday morning, the 25th, we heard the clocks of St. Paul de Leon at ſeven, being N. and S. of the Iſle de Bas; we hoiſted our flag, and fired a gun for a boat to come to us, which ſoon arrived, and carried Meſſrs. de Bougainville and l'Huillier de la Serre to Morlaix. The frigate proceeded to St. Malo. At eleven at night we dropt an anchor oppoſite the tower of Cape Frehel, the beacon being about a league N. W. $\frac{1}{4}$ W. of us.

On the 26th, at half paſt three in the morning, we got under ſail, and about ſeven came to our moorings in Solidor, where the veſſel was diſcharged. M. de Bougainville having given the King an account of our expedition, his Majeſty ratified the taking poſſeſſion of the Malouine Iſlands, and immediately iſſued orders for the Eagle to be got ready to return to theſe iſlands.

OBSERVATIONS
ON THE
STREIGHTS of MAGELLAN,
AND ON THE
PATAGONIANS.

THE King of France having approved of the poſſeſſion we had taken in his name of all the Malouine Iſlands, the miniſtry iſſued out orders for the ſupport and improvement of the little eſtabliſhment we had formed there. The Eagle frigate was again fitted out, and M. Alexander Duclos Guyot, who had been ſecond captain in the firſt voyage, was made firſt captain in the ſecond, with the brevet of lieutenant of a frigate, under the command of M. de Bougainville. Being informed of M. Duclos's ſafe return to St. Malo, and deſirous of knowing in what ſtate he had left the new colony, and what diſcoveries he might have made in the Streights of Magellan, I wrote to him to beg he would inform me of theſe particulars; and received the following anſwer:

" I waited to know what ſervice I ſhould be appointed to, before I did myſelf the honour of anſwering your letter; and therefore begin by acquainting you, that we ſhall ſet out from hence on the 10th or 15th inſtant, for the Malouine Iſlands. But I ſhall firſt put into the Madeiras, to take in wine, and other refreſhments. From thence I ſhall proceed to Port Deſire, on the Patagonian coaſt, to take a view of the country; and then go on to our colony, where after having unladen my proviſions, and landed all my paſſengers, I ſhall return to the Streights

Streights of Magellan, to fetch a cargo of wood; after which I ſhall wait for freſh orders from France. This is my deſtination: M. de la Gyraudais ſets out from Rochfort, with a frigate laden with proviſions neceſſary for the colony.

You deſire an account of my laſt voyage, which I ſhall now give you. We ſet out from St. Malo on the 5th of October 1765. We had in all on board the Eagle frigate 116 men, 53 of which were workmen, or officers going as paſſengers to the colony. Among the latter were M. de Perriers, a half-pay captain of the regiment of la Sare; M. Thibé de Belcourt, a half-pay captain of the regiment Dauphin; M. Denis de St. Simon, captain adjutant of the colonies; M. l'Huillier de la Serre, geographical engineer; M. de Romainville, lieutenant of infantry and engineer.

For the firſt fortnight, we had bad weather and contrary winds. On Sunday the 5th of November, we had a proſpect of the Cape Verd Iſlands. On Monday we paſſed by the iſlands of Fogo and Bravo. We did not meet with much ſtorm in croſſing the line, any more than under the tropics; and on Saturday the 16th of December, we came within ſight of the coaſt of Brazil, in 31 degrees 30 minutes South latitude, though we thought ourſelves, as it commonly happens, ſtill wide of it, from a defect in the charts, which you know throw this coaſt too far back weſtward.

We ſtayed afterwards ſome time ſearching for Pepy's iſland, where it is marked in the charts, and in the neighbourhood, without being able to find it. On Thurſday the 3d of January, we had a proſpect of the Malouine Iſlands, and made to land at the Iſlet, which in our firſt voyage we called *la Conchée*. On Saturday the fifth inſtant, we ſailed in, and caſt anchor at the diſtance of half a mile from the mouth of the ſmall bay of the colony; where we found every body in good health. We continued unlading till the firſt of February, when we were ready to ſet ſail. On Sunday the 27th, we had diſcovered three veſſels coming from the Weſt. On the 2d of February, ſeeing that they

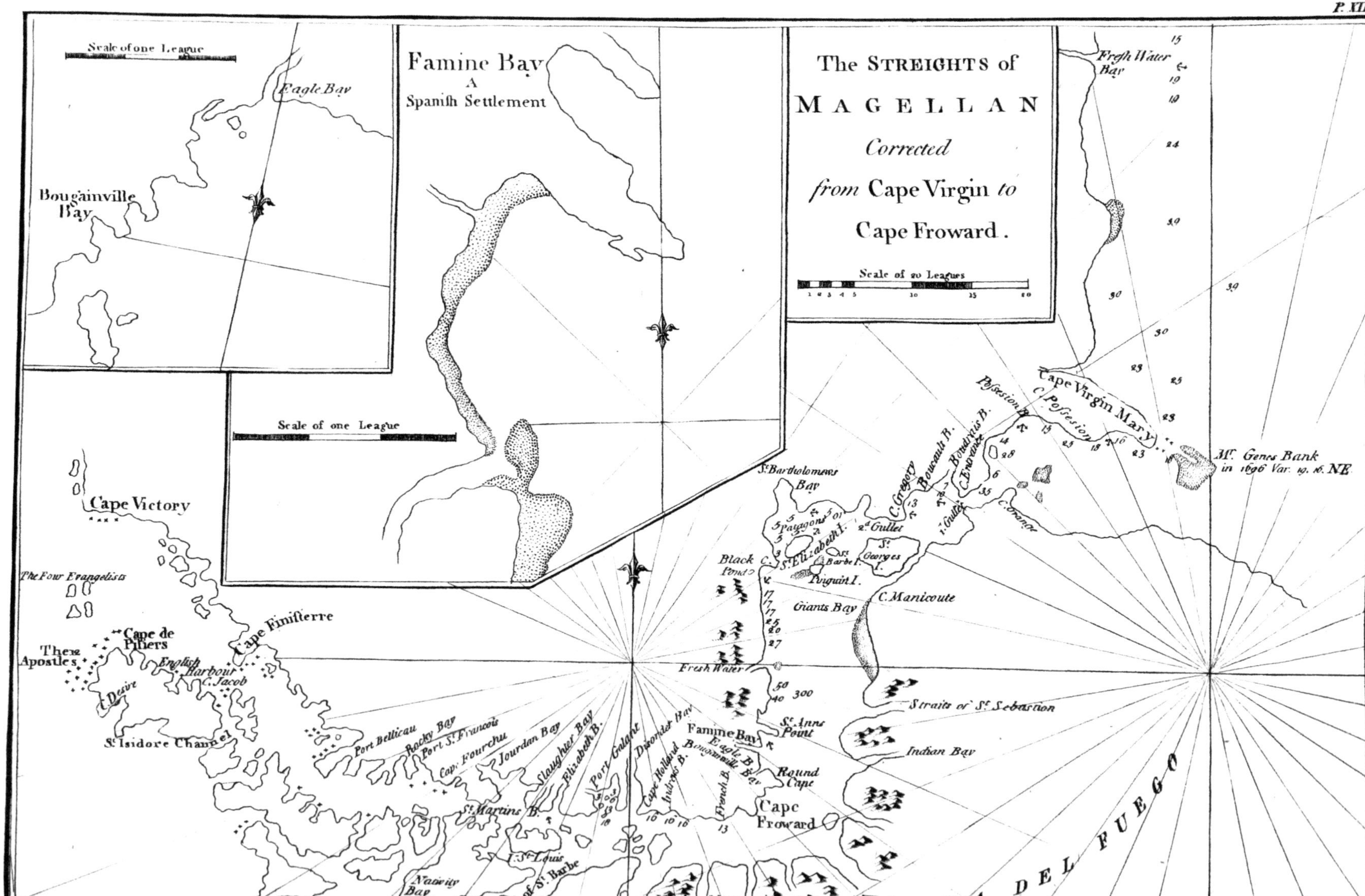
The STREIGHTS of
MAGELLAN
Corrected
from Cape Virgin to
Cape Froward.
Scale of 20 Leagues
Famine Bay
A
Spanish Settlement
Scale of one League
Scale of one League
Eagle Bay
Bougainville Bay
Fresh Water Bay
Cape Virgin Mary
C. Possesion
Possesion B.
Mr. Genes Bank
in 1696 Var. 19. 16. NE.
Boucault B.
Boudrais B.
C. Entrance
C. Gregory
1. Gullet
2d. Gullet
C. Orange
St. Bartholomews Bay
Patagons I.
St. Elizabeth I.
St. Barbe I.
St. Georges I.
Pinguin I.
Black C.
Black Pond
C. Manicoute
Giants Bay
Fresh Water
Straits of St. Sebastion
St. Anns Point
Famine Bay
Eagle B.
Bougainville Bay
Round Cape
Indian Bay
Cape Froward
French B.
Cape Holland
Andrews B.
Disorder Bay
Port Galant
Elizabeth B.
Slaughter Bay
Jourdan Bay
Cap. Fourchu
Port St. Francois
Rocky Bay
Port Bellicau
St. Martins B.
I. St. Louis
of St. Barbe
Nativity Bay
High Mountains
TERRA DEL FUEGO
Cape Finisterre
Cape Victory
The Four Evangelists
Cape de Pillers
English Harbour
C. Jacob
C. Desire
Three Apostles
St. Isidore Channel

they did not come into the bay, we ſet ſail for the Streights of Magellan. The weather was rather changeable during our paſſage. On Tueſday the 12th of February, we had a proſpect of Cape Lookout, on the Patagonian coaſt. After having tacked about, we found ourſelves within cannon ſhot of a lurking rock as large as our longboat, which we had a great deal of trouble to get clear of, on account of the currents, and the roughneſs of the ſea. This rock is not pointed out in our charts. On Saturday the 16th we obſerved three veſſels ſteering the ſame courſe that we did. On the 17th we entered the Streights of Magellan, together with the three ſhips. On Monday the 18th, one of the three ſhips working to windward while we were at anchor, ſhe ſtruck on a ſand bank. The weather was very fine. We ſent our boats to her aſſiſtance, with an officer, with anchors and cables; but ſhe ſoon diſengaged herſelf, and got off without injury. We then found out that they were Engliſh *. On

* This was in reality Commodore Byron's ſmall ſquadron. The fact is told in the printed account of his voyage round the world, in the following terms: " At four in the afternoon, the maſter of the ſtoreſhip *(the Florida)* came on board the Dolphin, bringing a packet from the Lords of the Admiralty to the Commodore.—He had likewiſe been ſeveral days in ſearch of Pepys's iſland, but was like us obliged to deſiſt.—To our great ſurprize in the morning of the ſecond day, after we left the harbour in company with the Tamer and ſtoreſhip, we diſcovered a ſtrange ſail, which indeed put us into no ſmall conſternation. The Commodore was inclined to believe, that this ſhip was a Spaniſh man of war of the line, who having got intelligence of our voyage, was come to intercept us; and in conſequence of that ſurmiſe, boldly gave orders, that all on board the Dolphin and Tamer ſhould prepare for a warm reception, by firing all our guns, and then boarding her from both ſhips; but while we were bringing to, and waiting for her, we found it grew dark, and we ſoon loſt ſight of her till the next morning, when we ſaw her at anchor, at three leagues diſtance, and therefore continued ſailing towards Port Famine. We however found that ſhe ſtill followed us, though at a great diſtance, and even came to an anchor when we did. On the 20th we were chiefly employed in getting up our guns; we ſoon got fourteen upon the deck, and then came to an anchor, having the Tamer aſtern, with a ſpring on our cable.

Thus buſily were we employed in taking all the meaſures prudence could ſuggeſt, to defend us from an imaginary danger; when an unlucky accident, which happened to the ſtoreſhip, ſhewed that we had nothing to fear, and that the veſſel, againſt which we were arming ourſelves, ought not to be conſidered as an

enemy;

On Wednefday the 20th, the Englifh anchored in Port Famine, and we failed on till the 21ft, when we caft anchor at the diftance of five leagues from the Englifh, and called the place the Eagle's bay, as it has no name on the charts. The next day, being the 22d, M. de Bougainville difcovering a very fine bay or port, at the diftance of one league and a half to the South, we went there and faftened the fhip to four trees very much under fhelter, at the diftance of a league from the French bay. We called it Bougainville bay. We took in fome very fine wood here and fhipped it conveniently, by hauling on board with a hawfer the wood cut upon the fhore. We ftayed here till the 16th of March, it being all the time very fine weather. On the 25th of February, two Englifh veffels, going to the South Seas, paffed by us. On the 16th of March in the morning, after having left a French flag, hoifted upon a hut, and feveral cloaths, kettles, hatchets, and other utenfils neceffary for the favages, we fet fail. After having gone a league a calm came on, and we caft our anchor in Eagle's bay. On the 17th it being calm, M. de Bougainville met fome of the favages as he was out a fhooting. He went up to them, and they appeared very gentle. On Tuefday the 19th we fet fail again; and on the 20th in the morning, the wind being againft us, we anchored in Port Famine. On the 21ft in the morning, fome of the favages calling out to us, we went up to them. They expreffed a great defire of coming on board: we therefore took fix of them along with us, whom we entertained, and who did not appear to be much furprized. They are a fet of men much like

enemy; for while the ftorefhip was working to the windward, fhe took the fhore on a bank about two leagues from our fhip. About the fame time, the ftrange fhip came up with her, caft anchor, and immediately began to hoift out her long-boats, to give her affiftance. But before they had come to the ftorefhip, our own boats had boarded her, and the commanding officer had received orders not to let them come on board, but to thank them in the politeft manner for their intended affiftance. We afterwards found this to be a French veffel; and having no guns that we could fee, fuppofed it to be a merchantman, who had come to thofe parts for wood and water.—On the 21ft we got into Port Famine, where we moored our fhips."

like the Indians of Montevideo, having no other drefs than the fkins of Sea-Wolves, Guanacoes, and Vicunas; they appear very poor, have no tafte for wine, but are very fond of fat. We dreffed them in red cloaths, and gave them feveral neceffary domeftic utenfils; we then accompanied them to land; crying out all the way *Vive le Roi de France*, which they repeated after us very well. We left a flag difplayed. They expreffed much good-will towards us, giving us their bows and arrows. When we faw them they were painted white, and in fpots, but as foon as we had given them fome red lead, not cinnaber vermillion, they immediately painted themfelves with it; and feemed to be fond of this colour. As we were returning to the fhip, they faluted us with *Vive le Roi* in French, having remembered that expreffion; and then they hollowed after their own manner, ftanding all round the flag. As we got farther from them, they raifed their fhouts, and increafed their fires.

This is nearly all I can tell you of thefe inhabitants of Patagonia. We did not land on the Terra del Fuego. I believe thefe are nearly the fame kind of people as thofe who crofs the Streights, in their canoes made of the bark of a tree. The firft time we faw them, they had kinds of hatchets; but they took care to conceal them afterwards, as well as their wives and children.

At length, on Saturday the 23d of March, we failed out of that famous ftreight fo much dreaded, after having experienced there, as well as in other places, that it was very fine and very warm; and that three-fourths of the time the fea was perfectly calm.

It is remarkable that the fea ebbs as it enters on the northern fide: we had a proof of this every day: in the middle the currents are diftinguifhable, but in the narroweft parts of the entrance they are very ftrong; they run at leaft two leagues and a half, and fink about four fathoms.

There is no wood at the entrance of the Streights, neither on one fide nor the other. There are nothing but immenfe plains.

About four and twenty leagues up the country, both on the coaſt of Patagonia, and on the Terra del Fuego, the woods begin. We found very little game, and that much followed by the natives, very little fiſh, and in the places where we had been, none of thoſe beautiful ſhell fiſh ſo much admired.

At length we ſteered our courſe in order to paſs to the South of the Danicant iſlands. On Tueſday the 26th, we came within ſight of land, which was the country to the Weſt of the Malouine Iſlands, about fourſcore leagues diſtant from Cape Virgin, which forms the entrance of the Streights. We afterwards ſailed fifty leagues to come back to caſt anchor in the port; ſo that we may reckon that we had paſſed by fifty leagues of the coaſt to the ſouthward, which is not however its greateſt length, as there is a ſtreight which divides the North and South lands without any woods. On the 29th of March we caſt anchor in the ſame place where we did before, having almoſt always had fine weather. We unladed our wood, and on the 27th of April we ſet ſail for France, leaving 79 perſons in the Malouine Iſlands. Our paſſage was rather tedious, by reaſon of the calm weather which obliged us to put into harbour, on account of the few proviſions we had remaining, having left as many as we could behind us. On the 18th of July we put into the harbour at Angra in the iſland of Tercera, where we ſupplied ourſelves with plenty of every thing, having found there all we wanted. On the 25th we left this harbour, and on the 13th of Auguſt arrived at St. Malo.

I am, SIR,

Your very humble Servant,

St. Malo,
1ſt September, 1765.

ALEXANDER DUCLOS GUYOT.

I was

I was not at Paris when M. de Bougainville returned, being gone to Montbriſon in Forez. He ſent me the following letter, which M. de Bougainville de Nerville his couſin, who was left commander at the Malouine Iſlands, had commiſſioned him to deliver to me. It was accompanied with a letter from him, which I ſhall give at the end of this.

If I had thought, Sir, you would have had the complaiſance to go and keep my mother company in her ſolitude, I ſhould not have omitted mentioning you to her, and deſiring that favour of you. She ſays ſo many handſome things of you, and you ſpeak ſo well of her, that I have reaſon to conclude you equally ſatisfied with each other. I am very glad to find that her company is ſo agreeable to you, and at the ſame time am much flattered in thinking myſelf ſometimes the ſubject of your converſation.

I ſhall now ſay ſomething to you about our ſituation. I have nothing particular to tell you about the winter we paſſed here. It has not been ſevere; for there never was ſnow enough to cover one's ſhoe-buckles, nor a ſufficient depth of ice to ſupport a ſtone as big as one's fiſt: and if it had not been for the rain, which runs through our tents as through a ſieve, we ſhould have had very little occaſion for fire, which we were now obliged to make in order to dry ourſelves. You would not have known our colony again had you returned with M. de Bougainville. In the firſt place you would have found us all very fat, the air being very healthy. You would have found all along the place where we live a fine walk of ſmooth even ground, and upwards of twenty feet wide; a new magazine raiſed again upon the border of the ſea; a fort completely repaired placed on a level, with platforms made with flat ſtones under the cannons; a new powder magazine, a bakehouſe, and a forge. By the account we kept, we killed above 1500 buſtards in the ſeaſon; for there is a time when they leave this country and go away to other parts, except a few ſtraggling pairs whoſe eggs we never could find; but only their young ones which were always ſix in number.

One brood of theſe was brought me, and was taken care of by one of our hens as her own. I was in hopes of ſending them to France, but ſince my couſin's arrival here, they have experienced a number of evils, and have at length all periſhed by the miſchievous tricks of the ſhip-boys who came to land: ſo that I muſt put this off till another ſeaſon. We have made the diſcovery of a bird much more beautiful than the buſtard, which is a kind of ſwan, as large and as white, but whoſe neck is as black as jet, and his bill red. We have not been able to kill any of them, as they are extremely wild. By other diſcoveries which I have made in the iſland more than twenty leagues to the Weſt, it appears that the part we dwell in is detached from other adjacent iſlands, or joined only by an Iſthmus. We may perhaps come at the true knowledge of this by means of the ſchooner which is to be left with us. The ſtudy of natural hiſtory, which we have not neglected, has furniſhed us with ſeveral of thoſe conchæ called *Poulettes*, or *Gueule de Raye*.

There are few of thoſe you ſent me the drawings of to be found in good preſervation. The Patellæ, you know, are common here. We have great expectation from our agriculture, our kitchen garden having ſucceeded very well. With regard to the corn, it produced in the dry land ſome beautiful ears; but they were fine only in appearance, having no grain within them. Our lands having been yet unſown require a longer time for cultivation, and muſt even be improved with good dung. We have not a ſufficient quantity of beaſts to make any trials with them. Four of our heifers and our three horſes are always in the open field; and we have never been able to catch them again, but their wandering diſpoſition has made us acquainted with one of the great advantages of this country: which is, that cattle may remain in all ſeaſons, day and night, in the open fields, without being in want of either paſture or litter. We often meet with one or other of them when we go out a ſhooting; they are as fat as hogs, and their liberty ſeems to agree very well with them. I return you many thanks for

the

the trouble you have taken in executing my commiſſions, and have received the things. I am making up a cheſt of the ſhells, ſeeds, and ſtones of this country; if you happen to be in the way when my couſin arrives he will ſhew them to you. They ſay you have put a ſhell into the cabinet of the Abbé of St. Germain, which is the only one of its kind. If that ſhell has been found here, be ſo kind as to ſend me a drawing of it.

From the Malouine Iſlands,
25th of April 1765.

DE NERVILLE.

The following is M. de Bougainville's letter.

I am at laſt returned, my dear fellow-traveller. At my arrival I found I ſhould not be able to ſee you, which gives me a great deal of concern. Be aſſured that no one intereſts himſelf more than I do in every thing that concerns you; and that I would have given any thing in the world, if you would have accompanied me in the ſecond voyage. We have made an alliance with the Patagonians, who have been ſo ill ſpoken of, and we have found them neither taller, nor even ſo wicked as other men. I ſend you a letter from my couſin, who has behaved admirably well. None of our people have ever been ſeized with a fever The winter has neither been ſevere nor long, and the eſtabliſhment ſucceeds very well. I brought them this year my ſhip full of the fineſt wood in the world, which I had from my friends the Patagonians. I have not at preſent time to enter into any more particulars, having not a moment to loſe. I believe I am going to be ſent into Spain, to ſettle ſome things with that Court relative to our new eſtabliſhment. I beg you would let me hear from you,

Paris,
uguſt 26, 1765.

and am, &c.

DE BOUGAINVILLE.

M. de

M. de Bougainville was ſent into Spain and ſettled matters between that Court and the Court of France, reſpecting the ceſſion which the latter made to the Spaniſh of the Malouine Iſlands; and M. de Bougainville ſet out from Nantz in 1766, on board a French frigate, and went to Buenos Ayres, taking a Spaniſh Governor from thence, and ſome troops of the ſame nation, to put them in poſſeſſion of the aforeſaid iſlands. Before he went away, he communicated to me the obſervations he had made on the Streights of Magellan, together with a correct chart of theſe Streights, which is among the plates, as well as a chart of the Eaſt, North, and South coaſts of the Malouine Iſlands, which they had paſſed by in going and coming back through theſe Streights. By this chart, one can judge only of the extent of the Malouine Iſlands to the North and South, the weſtern part not having yet been diſcovered. The Engliſh who ſettled themſelves in 1765 at Port Egmont, ſituated more to the Weſt than the French eſtabliſhment, may hereafter give us ſome informations with regard to thoſe parts yet unknown.

M. Alexander Duclos Guyot, and M. Chenard de la Gyraudais, having communicated to me the journals of their voyage they made together to the Streights of Magellan in 1766, with leave to make extracts from them; I have thought neceſſary to give theſe to the public, as well on account of the uſeful obſervations they contain relative to currents, the nature of the ſea, and the coaſts which form the Streights, as to ſettle the doubts of many learned men and others, upon the real exiſtence of the Patagonian giants.

Extract of the journal of M. Alexander Duclos Guyot, Lieutenant of a frigate, on board the Eagle frigate, in the Streights of Magellan, in 1766.

On the 24th of April, we ſet ſail from Acarron bay at the Malouine Iſlands. On the 26th, one of the Sebald iſlands that lies

lies moſt to the North Weſt, bore S. W. $\frac{1}{4}$ S. of us, diſtance 40 miles.

On the 28th in the morning, we ſaw a great quantity of whales and pinguins. At noon Cape las Barre as bore Weſt of us nine leagues.

On the firſt of May, at half an hour paſt ſeven, we ſteered W. S. W. in order to have a view of the land of Patagonia. At nine o'clock, bearing round the bank which is at the entrance of the Streights of Magellan, the ſea was changed, its waters being like thoſe of a river made muddy by rains.

On Saturday the 3d of May, at eight o'clock, Cape Virgin bore N. N. $\frac{1}{4}$ N. W. three leagues and a half, or four leagues. The moſt weſtern part of the Terra del Fuego S. W. $\frac{1}{4}$ S. Cape Santo Spirito S. S. E. Cape Poſſeſſion W. $\frac{1}{4}$ S. W. I think there are not leſs than ſeven leagues from one point of land to another at the entrance of the Streights.

On Sunday the 4th, at break of day, we were about four leagues S. E. of Cape Poſſeſſion. There is a ridge of rocks and a ſand bank near Cape Orange. It extends a great way, ſo we were obliged to coaſt the land of Patagonia. Here we ſaw a fire upon the ſhore, and drawing nearer to it perceived ſome men on horſeback, and many others on foot. When we came oppoſite to them, they called out to us, but we did not underſtand their language. We anſwered them with ſhouts, and hoiſted our flag. Five of them followed us about two leagues round the coaſt, but night coming on we loſt ſight of them. They ſeemed to be good horſemen, managing their horſes well, which were very active.

We hardly perceived any currents in the narrows, where it was almoſt a calm. This entrance in its narroweſt part is a full league over. At five o'clock in the evening, we anchored in Boucaut bay, in nine fathoms and a half water, with a bottom of rotten ſhells.

REMARKS on the TIDES.

IN the laſt voyage, I had obſerved, when we came into the firſt narrows, that the tide was coming in, and I reckoned it was the beginning of flood. I did not however perceive on the ſhore that the ſea roſe conſiderably; at which I was the more ſurpriſed, as all ſailors agree in ſaying that it does; neither was the ſhore wet, as it generally is when the ſea retires. On coming out we were two hours and three quarters making ſeven or eight knots, without getting on half a league. When the current diminiſhed, and we had ſailed half-way up the Streight, I perceived on its banks that the water had juſt fallen at leaſt four fathoms perpendicular. This obſervation induced me to imagine that when it is flood the ſea goes out on the northern ſide; but on the contrary, when it is ebb, it comes in, and bears to the South.

When we went along Cape Orange, we perceived a very extenſive flat ſandy ſhore, which we took for the open ſea at coming in, being concealed, as all the ridges and banks of Cape Orange are, which we could not ſee. This confirms me in my opinion, which is contrary to the ſentiments of all thoſe who have ſailed in theſe Streights before me. This day the tide was coming out, and was againſt us for ſome time; nevertheleſs the tide was very high when it began to bear to the South.

Then all the banks and ridges were concealed, as well as the flat ſtrands and ſhores which we had ſeen wet when we came out. I obſerved, that the tide bore in till nine o'clock. The ſea had then fallen four feet perpendicular: afterwards coming out again it roſe three fathoms; then there was a little interval without any ſtream, notwithſtanding which it ſtill roſe one fathom: afterwards the ſea took its courſe again, when it neither roſe nor fell while we went two-thirds of a league in an hour.

It

The Patagonians.

It afterwards fell without any current; which made me think the currents were not regular; and that in bays, the turn of the tide is caufed by the fwell. I fufpend the determination of this point, till it is confirmed by farther obfervations

We perceived about three o'clock in the afternoon, that the fea began to enter into the narrows, the Moon being 26 days old; which would make the fituation of the narrows E. and W. fo that it would be high water there at twelve minutes paft fix o'clock on the day of new and full moon.

On Tuefday the 6th, the favages appeared about nine o'clock in the morning, and were kindling a fire on the fhore by the fmall river Baudran. We hoifted our flag, and M. de la Gyraudais his broad pendant. Afterwards we both put our yawl and longboat to fea with men armed with mufkets and cutlaffes. In M. de la Gyraudais's longboat was an officer with prefents for the favages. In my yawl, we had feven failors and three officers under the command of my brother. At eleven o'clock we faw them land, and fome men on horfeback who received them; which appeared to me a good omen of peace. Nothing particular happened till twelve o clock.

My brothers account was, that the favages, who are natives of this country, were not the fame as thofe we faw laft year in Savage bay, and that they fpoke a different language. There were fix men and one woman who had but fix horfes, each guarded by a dog who never leaves them.

They received our people very well, coming up to them to fhew them where they fhould put into the fhore and land. They did not appear furprifed nor fhew the leaft fign of emotion. We meafured the fhorteft of them, and my brother found him 5 feet 7 inches high French meafure. The reft were confiderably taller. They were covered with the fkins of deer, guanacoes, vicunas, otters, and other animals. Their arms are round ftones, whofe ends are lengthened out and pointed. The round part is fixed to the end of a ftring compofed of feveral narrow ftraps, twifted and interwoven into a round form like the ftring of a clock, and making a kind of fling. At the other end of the ftring is another ftone in form of a pear, not more than half as big as the other, and appearing as if it was wrapt up in a bladder.

They ufe thefe weapons chiefly to catch animals; at which fport they are very dextrous, as they fhewed our people by an experiment made in their prefence. They have alfo other flings nearly of the fame kind of conftruction. They manage their horfes with great dexterity, and have a kind of faddle, very much like that we ufe for packhorfes. Thefe faddles are made with two pieces of wood, covered with leather and ftuffed with ftraw. The bit of the bridle is a fmall ftick, and the reins are twifted as the ftrings of their flings. They wear a kind of bufkins or half boots, of fkin with the fhag on, and two pieces of wood fitted to each fide of the heel, joined together in a point, which ferve them for fpurs. Their breeches are very fhort drawers, much refembling thofe of the favages of Canada, and are of a very good cut. It is evident they have had fome intercourfe with the Spaniards, from their having a very thin two edged knife, which they place between their legs. Their bufkins are made like thofe of the Indians of Chili. They pronounced fome words which were either Spanifh, or derived from that language. On pointing out the perfon who feemed tô be their chief, they called him *Capitan*. When they wanted to fmoke and afked for tobacco, they faid *Chupan*.

They fmoke in the fame manner as the inhabitants of Chili, throwing out the fmoke by their noftrils; and are extremely fond of a pipe. While they were fmoking they cried *Buenos*, ftriking themfelves upon the breaft.

We gave them fome new bread, and fome fea-bifcuit, which they eat with great appetite. The prefents we made them confifted in fome pounds of that red which we call vermilion: and fome red woollen caps, which however not one of them could put his head into: thefe caps though very large for heads of a

common

common ſize, were ſtill too ſmall for them. We alſo gave them ſome bedding, ſome hatchets, ſome kettles, and other utenſils.

My brother put his pocket-handkerchief round the neck of the chief; who having accepted of it, immediately looſened his girth, made of ſtraps twiſted together like the girth of a ſaddle, having at each extremity a ball of ſtone half incloſed with leather. There was alſo another ſtone fixt to the middle of the belt, and a whetſtone. He gave this belt to my brother, and faſtened it round his waiſt, expreſſing much friendſhip for him. We gave them to underſtand we were going on much farther in the Streights, and they made us comprehend by ſigns, that they would go to bed as ſoon as the ſun did, ſhewing us at the ſame time that they would lie down, and making a noiſe as if they were ſnoring in their ſleep.

As ſoon as our boats had quitted them and got out to ſea, they mounted on horſeback, and directed their courſe towards the place to which we had made them underſtand we were going.

They ſeem to be crafty, bold people, being more inclined to receive than to give. They wrap themſelves up in beaſts ſkins ſewed together, as the Spaniards do in their clokes. Our people killed ſome partridges; ſaw ſome wolves, foxes, and a great number of rats, but nothing curious.

On Thurſday at noon we caſt anchor under the low lands of Cape Gregory, in 25 fathoms of water.

After dinner we put our yawls to ſea to go a fiſhing and ſhooting. They came back in the evening without having taken or killed any thing, excepting one mangy vicuna, which M. Gyraudais ſhot. There are numbers of vicunas in this country, which is very beautiful. Our people ſaw a great quantity of foxes, wolves, and rats, and met with ſome few thickets of yellow wood, but no water.

On Friday the ninth we ſet ſail at day-break. At ten o'clock we got into the ſecond narrows, and ſteered our courſe in order to paſs between the iſlands of St. Elizabeth and St. Bartholomew. We afterwards anchored at 11 o'clock in the bay of Cape Noir, its

its point bearing N. N. W. 5 degrees N. where the wood begins to appear.

In viſiting the woods we found none but what was fit for fuel, and ſome yellow wood. The ſoil appears pretty good as well as the bay; into which we might have advanced much farther, the bottom being even: at 8 or 9 fathom of water, a fine ſand, and muddy nearer the land. In this bay one may be under ſhelter from the N. N. E. by the W.

We took in fiſhing only one large *Cornet*, ſome *Gras-dos*, with a golden fiſh, which was a kind of ſmelt. Our ſhooting parties were much leſs fortunate. By the great quantity of buſtard's excrement we found ſcattered about in the buſhes, we imagined, that that bird muſt be very plentiful here in the ſeaſon. There is no freſh water here; but there is a lake at the diſtance of a mile from the bottom of the bay.

On Saturday the 10th, at four in the morning, the ſea running eaſtward, fell twelve fathoms perpendicular. This appears contrary to all my fore-mentioned obſervations; but it might proceed from ſome croſs tide.

We continued along the coaſt of Patagonia, and found by our ſoundings the depth of water increaſe to 35 fathoms, muddy bottom, as we advanced towards the South. The coaſt is here alſo bordered with finer wood, which is found in greater quantities.

Having ſailed ſeven leagues in this direction, we came to the opening of a ſmall bay, where we met with a point, even with the ſurface of the water, and extending half a league out.

We had ſcarce ſailed one quarter of a league, after finding no ſoundings with a hundred fathoms, when all at once we met with no more than ſeventeen fathoms depth of water, and a little farther on, only eight fathoms, then five, then four and a half, with a fine ſandy muddy bottom. Soon after the depth increaſed to five and twenty fathoms. It is to be obſerved that it was high water. Perhaps the ſhalloweſt of theſe places would not have been covered at low water. There is no wood

wood upon this point, which is about ſeven leagues from Cape Noir; and the bank is one league South Eaſt of this point. This bank is not marked on the charts of the Streights; although it is very dangerous, being in the middle of the bay; which I imagine to be that called Freſhwater, by its diſtance from Port Famine. There are two ſmall rivers here, and ſome very fine wood; and it anſwers exactly the deſcription given of it by the Engliſhman, who named it Freſhwater bay.

We anchored ſoon after in Port Famine, which we ſounded, and found it good in every part. One may coaſt St. Ann's point at two cables length without danger, if forced to it by the wind; the leaſt depth of water there is five or ſix fathoms, which increaſes gradually to twenty-five at the diſtance of a quarter of a league: but one muſt not anchor here, becauſe the bottom is rocky, and there is a ſtrong current. At the South Weſt of this point there is a bank one cable's length from land, which is not three feet under water at low tide.

When the wind will allow of it, it is better to keep out a full mile from St. Ann's point, on account of the current; and leſt there ſhould ſtill be ſome lurking rocks under water, which may have eſcaped our notice: one may caſt anchor in eight or ten fathoms water, rather towards St. Ann's point, than towards the South; for here the depth of the ſea ſuddenly diminiſhes even at high water, as well as in the bottom, where, at low water, there appears a ſhallow, uncovered for more than a quarter of a league.

On Saturday the 17th, we ſounded the ſmall bays to the North of St. Ann's point, where we found ſome banks, extending far out.

On Sunday the 18th, we ſent after dinner all our carpenters on ſhore, to cut ſome wood for burning and building; which was the reaſon of our being ſent here, as well as to fetch away ſome trees for planting.

On Wedneſday the 28th, M. de la Gyraudais being laden and ready, ſet ſail at ſeven in the morning to return to the Malouine Iſlands.

On Friday the 30th, in the morning, I perceived ſome ſavages upon the ſandy iſland, which forms the South entrance of the bay where we had left them the year before. I went to them, and knew them to be the ſame ſavages. They were two and twenty men, without women or boats. Having no preſents to give them, and not being able to make them comprehend me, I embarked again.

On Sunday the 1ſt of June, early in the morning, the ſavages made ſome ſigns to us; but the bad weather prevented us from coming to them. They made us underſtand, that they wiſhed we ſhould get into the river with our yawl.

On the 2d, two of the ſavages appeared at the bottom of the bay, calling out to us in their language. I ſent an officer in the yawl, to aſk them if they would come on board. On the arrival of the yawl they fled towards the river beckoning us to follow them. The officer thought it more prudent not to do it, and came on board again. At eleven o'clock we ſaw them come out again in ſix canoes. They croſſed the bay, paſſing within muſket-ſhot of us, but would not come on board: they went and landed in a little creek under St. Ann's point. As I had put ſix men in this place to cut fire-wood, and the ſavages were very numerous, I immediately armed the yawl and the longboat, and went to meet them. At my arrival, ſome of them were employed in building their huts; others were fiſhing for ſhell-fiſh, muſcles, patellæ, ſea-urchins, crabs, bucinna, taking all theſe only from the rocks. Notwithſtanding this they have nets made with cat-gut.

After having renewed the alliance made laſt year, I diſtributed preſents among them, conſiſting in ſome pounds of vermilion, ſome woollen bed-cloaths, ſmall looking-glaſſes, chalk, knives, ſome clokes, a hatchet, bread, &c. They would not taſte any wine. I did not chuſe to offer them brandy, leſt their acceptance of it might be attended with dangerous conſequences.

Their company appeared to conſiſt of twenty ſix men or boys, and forty women and girls, among whom were a great number

of

of young people. The Chief of them is called Pacha-chui. He is diſtinguiſhed from the reſt by a cap of birds ſkins with the feathers on. When he receives any viſits he puts it on his head, which is, no doubt, meant as a mark of his dignity. The preſence of the men, who ſeemed exceſſively jealous, obliged the women to aſſume an appearance of great modeſty.

I queſtioned the Chief as well as I could about his religion. He gave me to underſtand, at leaſt I thought I comprehended by his ſigns, that they neither worſhip the ſun, moon, men, nor animals, but only the heavens or the whole univerſe; this he repeated ſeveral times, always lifting up his hands joined together over his head.

During this time they continued throwing upon the fire, without any ceremony, all the wood cut down by our people. This obliged me to ſend my ſix men to cut wood at a greater diſtance from theſe ſavages, to avoid quarrelling with them.

They exchanged with our people ſome bows and arrows, and ſome necklaces of ſhells in return for cloaths. I then left them, and invited them to come on board. Four of them accepted my invitation. I made them dine with me, and entertained them in the beſt manner I could. They preferred bacon to every thing elſe. Their deſert was a candle to each, which they devoured with great eagerneſs. When dinner was over, I had them dreſſed from head to foot, and gave them ſome trifles with which they appeared very much pleaſed; and then ſent them to land.

In the afternoon I returned to the huts of the ſavages. The Pacha chui came to meet me, and made me a preſent of a kind of flint to ſtrike fire, like thoſe which are found in Canada, appearing to be a marcaſite of yellow copper. He afterwards diſtributed the preſents I had made them in the morning.

One of them was continually muttering; I aſked him the reaſon of this. He gave me to underſtand that he was ſaying his prayers, by pointing up to the heavens as the Pacha-chui had done in the morning. This ſeemed to imply that they worſhipped ſome divinity, but I could not comprehend what the divinity was, nor under what title he was adored.

Both men and women have no other dreſs than the ſkins of ſea-wolves, vicunas, guanacoes, otters, and lynxes, which they throw on their ſhoulders. Moſt of them are bare-headed. A bird's ſkin with the feathers on, covers their private parts. The men call themſelves *Pach-pachevé*; the women *Cap-cap.* They taught me theſe names by ſhewing me firſt their perſons, and afterwards the parts which diſtinguiſh the ſex. Both men and women are thin. Their canoes are ill-built, in compariſon with thoſe of the ſavages of Canada. The women are the perſons employed in rowing and fiſhing. They have a number of dogs, reſembling foxes; which they call *Ouchi*; and their canoes, *Shorou.*

It is to be obſerved, that the morning tides are always equal every morning; riſe very little in open ſea, and are only as the neap-tides.

On Wedneſday morning the 4th, the ſavages made no ſcruple of burning five or ſix cords of wood, which our people had cut down, but they aſſiſted in bringing the reſt on board.

At noon the Pacha-chui came on board our frigate, attended by eleven men. I made him dine with me, and gave the others ſome biſcuit, and a piece of tallow; and for their drink three pints of the oil of ſea-wolves. They ate and drank all up with a moſt excellent appetite. I afterwards dreſſed the Pacha-chui, and giving ſome trifles to the others, ſent them all on ſhore.

On the 6th, all the ſavages, pleaſed with the reception I had given their comrades, came in four canoes to pay me a viſit. But as they had large fires in their canoes, I would not ſuffer them to come on board, at which they ſeemed diſpleaſed. I ordered them ſome biſcuit and oil; and after dinner ſent them back without giving them any reaſon for it.

On Sunday the 8th, the ſavages began to be troubleſome: they ſtole ſeveral hatchets, ſome proviſions, and cloaths from us. As they ſeemed inclinable to theft and fraud, I took the reſolution to let nobody lie on ſhore, and to ſubmit to the inconvenience of having all the utenſils and tools brought back every night.

On Monday the 9th, the ſavages ſtole again ſome harpoons, hatchets, iron-wedges, and mauls. I complained of this to the Pacha-chui, and deſired our tools might be returned, but to no purpoſe. I then gave them to underſtand, that if they perſiſted in theſe practices, we ſhould treat them in a different manner.

Their boat, which had croſſed the bay laſt night, now brought them a dying man, about forty years of age, who was exceedingly emaciated.

In the afternoon, our wood-cutters repreſented to me, that they loſt a great deal of time in coming back to lie on board, and then returning in the morning to the wood: they therefore aſked leave to lie on ſhore. I conſented to this, deſiring them at the ſame time, to treat the ſavages mildly if they came to viſit them.

For this purpoſe, I placed a diſcreet perſon at the head of them, and with him his brother, a man of a mild diſpoſition, and who, from being uſed to live among the ſavages of Canada, was in ſome meaſure acquainted with their manners: and after recommending it to them to keep a ſtrict watch leſt they ſhould be ſurprized, I returned to the ſhip.

On Thurſday the 12th, about four o'clock in the morning, we heard ſome noiſe among the ſavages. Three of their canoes, with a great number of women in them and ſome men, came up to our frigate. I gave them ſome pieces of bread, and ſome oil of ſea-wolves, the greateſt part of which they put into a kind of bladder they had brought on purpoſe, and drank off the reſt. I would not ſuffer them to come on board, on account of their being ſo much addicted to theft, and becauſe they had got large fires in their canoes. This day I obſerved, contrary to the common cuſtom, that the men were not painted: only ſome few of them were painted black, which gave them a very frightful appearance. The women were all ſpotted with black, having their faces and necks bloody, as if they had ſcratched themſelves with thorns. Two of their canoes doubled St. Ann's point going to the North.

On Sunday the 15th, in the morning, I went to pay a viſit to the ſavages. Not ſeeing the ſick man, I aſked them what was become of him; they made me comprehend he was dead. The cries we had heard on Thurſday morning were probably the marks of their mourning. They ſeemed all very much afflicted, and were all painted black, contrary to the uſual cuſtom; and the women appeared ſcratched all over, as if they had been torn with pins. I obſerved that they ſhewed much regret for the dead man. I aſked them by ſigns what they had done with him. They anſwered me only by lifting up their hands to heaven, repeating the ſame ſigns ſeveral times, in order I ſuppoſe to make me underſtand the deceaſed was there: from whence it may be conjectured that they believe in a future ſtate. They would never tell me what they had done with the dead body. I am inclined to think they had tranſported it in one of their canoes, with which they had doubled St. Ann's point. I diſtributed ſome biſcuit and oil of ſea-wolves among them.

On Monday the 16th, I perceived two canoes of ſavages coming towards us, and all the reſt going out of the bay. I put myſelf into our yawl, taking ſome bread and oil along with me. When I came near them, I made them a ſign to follow me to land which they did very readily. I gave them the bread and oil. They broke up their camp, and thoſe who ſtaid behind were gathering up the remains of it. They made me underſtand that they were going to live at the diſtance of a league from that place, in one of the ſmall bays to the North of St. Ann's point, becauſe the ſhell-fiſh became ſcarce in the place where they were. The Pacha-chui was in one of the two boats, and was coming with an intention to thank me, and to apprize me of his departure.

I then ventured to aſk him, if any of his young people would come away with us, making him underſtand as well as I could, that I would bring him back in a twelvemonth. He anſwered by by ſigns that he conſented, and immediately preſented one of them to me, who ſeemed ſatisfied. We then left each other, and I brought away my young ſavage with me, to put him on board. I dreſſed him, and entertained him as well as I could. The Chief went out of the bay to join his troop.

On Tueſday the 17th our ſavage ſeemed to be pleaſed with us; and even looked contented and chearful. About ten o'clock, ſeventeen ſavages coming by land from a ſmall bay which lay North of us, and where they were encamped, paid a viſit to their companion. We went to meet them, taking him along with us; and I gave them ſome bread and oil for their breakfaſt. As we were going back, another of them aſked leave to come on board to ſtay with his comrade. As the offer was voluntary I took him along with me.

Towards ſix o'clock in the evening, I perceived that our two ſavages were ſo melancholy as even to ſhed tears, and that they were conſtantly looking towards land. I was not at a loſs to find out the cauſe of this uneaſineſs; and thought it natural that they muſt on reflection regret the reſolution they had taken. Notwithſtanding my deſire of bringing them away, in hopes that I might afterwards receive ſome uſeful information from them, I determined to ſend them back, and reſtore them to that liberty which they certainly imagined they had loſt. I made them get into our yawl, and had them conducted back to land. They expreſſed much joy when they came on ſhore, and deſired they might go to their families.

On Wedneſday, at 9 o'clock, they came to aſk for ſome bread and oil. I ordered ſome to be diſtributed to them, and having aſſiſted in loading our longboat, they went back to their firſt encampment. At four o'clock in the afternoon they left us, making me underſtand that they were going to reſt, becauſe the moon, which they call *Sercon* was up; but that they would come beak, and bring with them the two young men who had been on board of us. When we got back to our ſhip, we heard two guns fire; the ſignal agreed upon between us to call for help, in caſe

caſe we ſhould be attacked by the ſavages. I then ſuſpected that our people were engaged with them. I immediately had our boats armed, and ſent them to their aſſiſtance, but it was too late: the victory was already gained, and the ſavages routed when we landed. The affair happened in the following manner:

Twenty, or ſix and twenty ſavages, as we were told, came down ſecretly and ſilently through the wood behind the work-ſhop; and three of them entered ſuddenly into the hut where our people were, who thinking that the ſavages ſeemed to have ſome miſchievous deſign, placed themſelves at the entrance of the hut to hinder the reſt from coming in. They then attempted to force their way, and not ſucceeding fell upon our men, ſome attempting to ſeize their legs, in order to throw them down, and probably to bind them, being provided with large ſtraps in form of ſlings, having at the end a dart about ſix inches long, made of a jagged bone; the reſt beat them with large ſticks. Our people, though they were ſurprized at ſo ſudden a declaration of war, were not diſcouraged. They ſeized their cutlaſſes, and exerted themſelves bravely againſt their enemies, deſtroying as many of them as they could; by which means they threw the ſavages into confuſion and routed them: our people however were but ſeven againſt twenty-five: three ſavages remained dead upon the field of battle, excluſive of the wounded; three of our people were wounded; the maſter carpenter received ſeveral blows upon the head with a ſtick; another was dangerouſly wounded in the head with a cutlaſs; and his brother was cut on the hand with the ſame inſtrument, which has quite diſabled him. The wounded were dreſſed as ſoon as they came on board. One of the three was afterwards trepanned.

On Friday the 20th, in the morning, I ſent the longboat to fetch away the timber, and to bury the three ſavages in the ſame grave. After having raiſed the ground to a certain height, we placed their ſkins or cloaks, with their ſhoes on the top, that the other ſavages might find out the place where their dead companions lay; and that they ſhould not think we had eaten them;

which

which perhaps they might do, if they were unable to find the dead bodies.

On Sunday the 22d, we were at the entrance of the narrows; and at eleven o'clock we ſaw ſeveral fires on the low lands of Cape Gregory. In coaſting theſe, we diſcovered about 90 or 100 men, moſt of them on horſeback, who followed us to the place of anchorage; thinking, without doubt, that we ſhould anchor there. But I was prevented by the wind blowing freſh, and the weather being favourable for ſailing out of the Streights. We made twelve leagues ſince morning, the ſavages making ſigns to us all the while. At nine o'clock in the evening we cleared Cape Virgin, and left the Streights.

Remarks made in 1766 in the Streights of Magellan, from Cape Virgin to Cape Rond; by M. de la Gyraudais, Lieutenant of a Frigate, at preſent Captain of a fire-ſhip.

CAPE Virgin is of the ſame height as Cape Fréhel, in the road of St. Malo, and has the ſame form. At two leagues and a half weſtward, it ſends forth a low point, which extends a league out at ſea to the South, with a ridge of rocks, which is covered by the tide at two cables length from this point; and againſt which the ſea breaks with great violence. This ridge is not marked upon the chart of the Streights, any more than a bay in which we anchored. The coaſt is rather high and ſound, from Cape Virgin to Cape Poſſeſſion. One may ſail along it at the diſtance of half a league without any danger. Poſſeſſion bay is large. It ſhelters ſhips from the wind, from the W. S. W. to the N. E. paſſing by the N. It may eaſily be known by M. de Gennes' plan, which is accurate in the diſtances, and the bearing of the lands; except with regard to Lion's iſland, which he does not place ſufficiently to the W. S. W. by one league and a half at leaſt. Over Poſſeſſion bay, there is a large cape, and to to the S. W. of this, four ſmall hummocks ſituated near each other.

From this bay till one gets beyond the firſt narrows, the coaſt is low and ſound on the ſtarboard ſide going in. After this comes the bay Boucaut, formed by the firſt narrows, and Cape Gregory which is pretty high. Two leagues in land there is a mountain, running N. E. and S. W. and a very high even land, which is ſeen a long while before one enters the firſt narrows.

After paſſing the ſecond narrows, the land riſes, and there are ſeveral hollows from this to St. Elizabeth's iſland; and from thence to the main land, which muſt be coaſted as near as poſſible, particularly on the flood; becauſe the tide throws with prodigious force againſt St. Bartholomew's iſland. The paſſage is between theſe two iſlands to Cape Noir, which is high; and where there is very convenient and good anchorage, called by M. de Gennes, Freſhwater, though it is not ſo. Here we begin to find wood. Freſhwater is ſix leagues beyond, in a creek, the ſtarboard point of which is very low, and where there is no wood. Here in ſounding acroſs we had no ground at fifty fathoms. Two minutes after we ſaw the bottom, and at four fathoms fine grey ſand. We followed this bottom a quarter of a league farther, taking care to keep rather wide of it. I would adviſe not to go nearer to it than the diſtance of two leagues. From thence to Port Famine the land is high, as it is alſo to the bay of Cape Rond.

Obſervations on the Terra del Fuego ſide, from the entrance of the Streights.

FROM the ſide of Cape Virgin, to two leagues and a half within, the land is high and ſound. Here there is a very low point, extending one league out at ſea S. E. and N. W. To the N. and S. of this point, and one league wide of it, is a ſhoal bottom.

bottom. The coaſt afterwards forms a hollow, not to be ſeen but in fine weather, as far as Cape Orange, which makes the entrance on the laiboard ſide of the firſt narrows. Here is a ridge of rocks which covers at high water, and extends N. E. S. W. to the diſtance of two long leagues from this cape. From hence to the ſide of the ſecond narrows, the land forms another hollow; and from the ſecond narrows to the ſide of Cape Rond, the land is very high, and forms an appearance of four hills. Between theſe poſſibly there may be ſome bays. M. de Gennes has not marked the two low points of land, placed before, and on this ſide of Cape Rond, at one league and a half or two leagues diſtance.

Extract of the Journal of the ſame M. de la Gyraudais commanding his Majeſty's Pink, l'Etoile, going from the Malouine Iſlands to the Streights of Magellan.

I Believe there is a greater diſtance between the Malouine Iſlands and the main land of Patagonia, than is marked upon the charts; for the Eagle found herſelf by her reckoning eighteen leagues ahead of the ſhip, as well in going as in coming back. We ſounded frequently, and found ſixty fathoms, mixed bottom, with white coral, and a gun flint, which was remarkable enough. Here we ſaw ſeveral whales, ſome ſea-larks, larger than common, ſome pinguins, divers, petterils, oſpreys, and large gulls.

From the twentieth to the firſt of May we had thick weather, which hindered us from ſeeing land, when we were more than half a league diſtance from it: the ſea ſeemed agitated as if we had been in a race. The water was here diſcoloured at eight leagues from the ſhore, but more ſo at the mouth of the Streights. At ten, the weather clearing up, we ſaw land, diſtance four leagues. By our reckoning we still found the Malouine Islands farther off from the main land, than is lain down by our charts.

On the fifth instant, about four in the afternoon, we saw a fire on the coast of Patagonia. Upon coming nearer, we saw seven men with their horses. We could not discern whether they were naked or clothed. When they perceived that we had got beyond the place where they had made their fires, they followed along the coast, mounted upon their horses, and dogs after them. Seeing that we continued our course, they shouted, but we could not comprehend their meaning. The wind and tide being in our favour, we lost sight of the Patagonians and passed the first narrows. It was a league and a half over. Between five and six we anchored in the bay Boucaut, at three leagues from Cape Gregory, with ten fathoms water, muddy bottom of sand and small shells, at the distance of a full league from the land. One should not cast anchor in lesser depth of water; for the sea fell three or four fathoms in the night-time. The coast is well laid down in M. de Gennes' plan.

From the 6th to the 7th, in the night, we again saw fires on the Patagonian coast. At eight o'clock this fire was of one side of us, and we distinguished some Patagonians on shore, by means of our spying glasses. The Eagle and myself put out our yawls to sea, and sent them with fifteen men well armed, including the officer, to the spot where we saw seven of the savages. They paid our people some compliment in their own language. Our seamen could not understand them; but imagined their faces and behaviour expressed a satisfaction at seeing us. After the first compliments, they conducted our people to their fires.

Here they examined the Patagonians at their leisure, and found them to be men of the highest stature: the least of them was five feet seven inches *(French measure)*, and of a bulk beyond the proportion of their height, which made them appear less tall than they are. They have large strong limbs, and broad faces; their complexion is extremely tanned, their forehead high, their

their noſe flat and broad; their cheeks are full, and their mouth large; their teeth are very white, and well ranged, and their hair black. They are ſtronger than our Europeans of the ſame ſize.

The words they pronounced were, *Echoura*, *Chaoa*, *Didon*, *abi*, *abi*, *obi*, *Choven*, *Quécallé*, *Machan*, *Naticon*, *Pito*. Theſe were the only words our people could gather, while they were warming themſelves at their fires.

M. de St. Simon, an officer, who by order of the miniſtry embarked with us for the Malouine Iſlands with preſents for the natives, acquitted himſelf extremely well of his commiſſion. He gave them ſome harpoons, bludgeons, bedding, woollen caps, vermilion, and in ſhort every thing he thought would be moſt agreeable to them. They appeared very well pleaſed.

They are clothed with the ſkins of guanacos, vicunas, and other animals, ſewed together in form of ſquare clokes which reach below the calf of the leg almoſt to the ancle. They have a ſort of buſkins or half-boots, made of the ſame ſkins, with the ſhag on the inſide, as it is alſo in their clokes, which are very well ſewed together in regular compartments, and painted on the outſide with blue and red figures, bearing a reſemblance to Chineſe characters. The figures however are almoſt all alike, and divided by ſtraight lines which form ſorts of ſquares and lozenges *. They have ſomething like hats ornamented with feathers, much in the ſame manner as ours. Some of theſe hats reſemble very much the Spaniſh caps.

Several of our people went a ſhooting at ſome diſtance, where they killed a few partridges, and ſaw ſome carcaſes of vicunas.

 The

* M. de la Gyraudais received as a preſent from theſe Patagonians, when he viſited them at his return to the Malouine Iſlands, ſeveral of their clokes, ſome of their weapons, ſome ſlings armed with ſtones, and ſome necklaces of ſhells from their women. He brought them to Paris, and gave part of them to M. d'Arboulin, who had ſome of them preſented to the King, and kept the reſt. I examined them at leiſure, and although I am rather more than five feet ſeven inches (French meaſure) one of theſe clokes thrown on my ſhoulders, (as the Patagonians wear them) trailed on the ground at leaſt a foot and a half.

The country they went over is uncultivated, barren, and dry There is nothing but heath upon it, and very little graſs. The horſes of the ſavages ſeem to be very bad but they manage them with great dexterity. The Patagonians made ſome preſents to our people who were returned from ſhooting. Theſe were round ſtones, of the ſize of a two-pounder ball. They are placed in a ſtrap of leather, faſtened and ſewed to the end of a ſtring of catgut twiſted like a rope. It is a kind of a ſling, which they uſe very dexterouſly for killing animals a hunting. On the end, oppoſite to that which fixes the round ſtone, there is another ſtone placed, half the ſize of the former, and cloſely covered all over with a kind of bladder. They hold the ſmall ſtone in their hand after having paſſed the cord between their fingers; and then making a turn with the arm, as in caſting a ſling, they throw the weapon at the animal, whom they can reach, and kill at the diſtance of four hundred feet.

The complexion of the women is tolerably clear, for they are much leſs tanned than the men, yet they are proportioned to them in ſize. They are alſo dreſſed in a cloke, wear buſkins, and a kind of ſmall apron, which only hangs down half the length of their thighs. They certainly pluck out their eyebrows for they have none. Their hair is dreſſed in front, and they have no hats.

Theſe Patagonians are ignorant of the paſſion of jealouſy, at leaſt there is reaſon to think ſo, from their encouraging our people to handle the breaſts of their wives and daughters, and making them lie promiſcuouſly with them, when I paid them a viſit on my return to the Malouine Iſlands.

We gave them bread which they ate, and ſome tobacco for chewing and ſmoking. By their manner of uſing it, we ſaw plainly it was no novelty to them. They would not drink any wine. When we had been five or ſix hours with them, they grew more familiarized. They were very curious, ſearched our pockets, were very deſirous of ſeeing every thing, and examined us with attention from head to foot.

We

We mounted their horfes, which were equipped with bridle, faddle and ftirrups. They ufe both whip and fpurs; and feemed fatisfied and well pleafed to fee our people ride their horfes. When I had a gun fired for fignal to bring our people back, they fhewed not the leaft emotion or furprife. When we went away they entreated us much to ftay with them, giving us to underftand by figns, that they would fupply us with food, and though they had nothing to offer us at prefent, yet they foon expected fome of their people to return from fporting. We anfwered them alfo by figns that we could not poffibly ftay; and that we were going directly to a certain place, which we attempted to point out to them, endeavouring at the fame time to make them comprehend that we wifhed them to bring us fome oxen and horfes. We know not whether they underftood us.

On the eighth, having fet fail from Bay Boucaut, and anchored under Cape Gregory, we went a fhooting on fhore, and the foil appeared the fame as on the laft fpot. After we had walked about a league, we met with two herds of vicunas, each confifting of three or four hundred, of which we could not kill more than one with a mufket charged with ball. I alfo fhot a *Stinkbingfem*, which I left on account of its offenfive fmell. I likewife fired at a wolf, but all thefe animals are very wild, and will not fuffer any one to approach them.

At half paft fix in the morning of the ninth, we got under fail in very pleafant weather. M. de Gennes in his draught lays down the fecond narrows Eaft and Weft corrected by the globe, but he has marked it two points too much to the Weft. I would advife to keep the Patagonian fhore till you come to the North and South of Elizabeth's ifland; on account of the ftrong tide which runs upon St. Bartholomew and Lyon iflands, and upon fome fhoals lying off thofe iflands. We coafted clofe to Elizabeth s ifland, till we came to Cape Noir, where we anchored in eight fathoms water, fandy and muddy bottom, with broken fhells.

From

From Friday the ninth to the tenth, we kept along the Patagonian ſhore, at the diſtance of a league and an half. The coaſt appeared woody, but on the return of our boat we were told the wood was not good for much. Being near a low point we ſounded, and no ground at fifty fathoms. An inſtant after we ſaw the bottom, which was ſandy, and at four fathoms water; this obliged us to haul off.

From the 10th to the 11th we had much wind, and foggy weather, with a very rough ſea. As we were no more than five leagues from Port Famine, I determined to go and anchor there. The Eagle followed us, and we ſoon had reaſon to be pleaſed with this reſolution, for a quarter of an hour after we came to anchor, we could not diſcern any object at the diſtance of half a cannon-ſhot from us, and the wind ſtill continued blowing very hard.

From the 11th to the 12th, the fog and rainy weather continued. Having walked round the bay, we met with ſome fine wood, and diſcovered a very rapid river, on the larboard point of the mouth of the bay. This ſtream makes the ſea as dirty and as turbid, as a river overflowing from abundance of rains.

On the water ſide there were ſeven or eight huts belonging to the ſavages, which they had but lately quitted. I fired a gun, and hoiſted our flag, in order to attract the ſavages from the neighbouring parts.

From the 13th to the 14th there was a high wind, followed by a prodigious violent ſtorm, which ended in a great fall of rain, ſucceeded by ſnow and hail, which laſted till noon, when the weather grew calm.

From the 16th to the 17th, we met with ſome very fine wood, and ſent an officer and thirty men on ſhore, to pitch a tent, and cut roads through the woods. We were conſtantly employed in cutting and ſhipping our wood till the 17th, when we unmoored, leaving the Eagle to complete her cargo, and bring up ours to the Malouine Iſlands.

From the 29th to the 30th, at ten in the morning, we ſaw a fire on ſhore, which the ſavages had kindled on our account. We ſteered towards the fire, and ſaw ſome men and horſes.

From the 30th to the 31ſt, the night coming upon us unawares, we came to our anchorage by the light of two fires which the ſavages had made for us, one upon a mountain, the other upon the ſea-ſide. We anchored in nineteen fathoms, black muddy bottom, with ſmall ſhells.

At day-break the ſavages ſhouted, in order that we ſhould come to them. I put my yawl and longboat to ſea well armed, and with preſents. I went on ſhore, where I found three hundred ſavages, including men, women, and children. Not expecting to meet with ſo many, I was obliged to go on board again to fetch ſome more preſents.

From the 31ſt to Sunday the firſt of June 1766, the wind having driven our yawl from ſhore, which was empty, our people were under ſome anxiety for fear of loſing it. The ſavages perceiving this, one of them who was on horſeback, ſpurred his horſe, and plunged with him into the ſea, to ſwim after the yawl. He got hold of it, and brought it back to our ſeamen Perhaps we who pique ourſelves ſo much upon our politeneſs, affability, and humanity, and who call theſe Patagonians ſavages, would hardly have done ſo much for them, in a ſimilar circumſtance.

At ſeven in the morning the longboat went to ſhore with the reſt of the preſents, which the ſtormy weather had prevented us ſending ſooner. It came back with thirteen of our people who had ſtayed with the ſavages ſince yeſterday morning. They told us that theſe Patagonian giants had treated them with the utmoſt civility according to their manner, and given them marks of the ſincereſt friendſhip, even ſo far as to invite them to lie with their wives and daughters; that they had given them ſome fleſh of the guanacos, ſeveral of their clokes, and ſome of their ſlings; and the women ſome of their necklaces made of ſhells.

They

They alſo made me a preſent of twelve horſes; which I could not keep for want of forage.

The piece of civility moſt troubleſome to our folks, was that of being obliged to lie promiſcuouſly among the Patagonians; who often lay three or four together upon one of our people, to keep the cold from them; ſo that their muſkets and other arms became uſeleſs. They would therefore have had no reſource left but in their pocket-knives, which would not have been of much ſervice for defending them, in caſe of neceſſity againſt five or ſix hundred men, including women and children, and all of them proportionally of an enormous ſtature, both in height and bulk. Each man or woman, had one or two dogs, and as many horſes. They ſeemed to be of a mild diſpoſition, and very humane. It would be eaſy to eſtabliſh a very profitable trade with them, for their horſes, and for the ſkins of vicunas, which are ſo much valued, and bear ſo high a price in Europe. The ſkins of guanacos are alſo excellent, though not ſo fine.

From the 7th to the 8th, a very high wind, rainy and thick weather. The ſea was terrible, the wind blowing always by ſqualls.

From Sunday the 8th to the 9th, the ſea was very rough, with rain, hail, ſnow, and fogs. At nine we ſaw land without knowing what it was: at noon we found it to be Sebald de Wertz iſlands, which bore S. E. diſtance ten leagues.

On the 15th we caſt anchor in Acarron bay, in the ſame place from whence we ſet out.

FINIS.

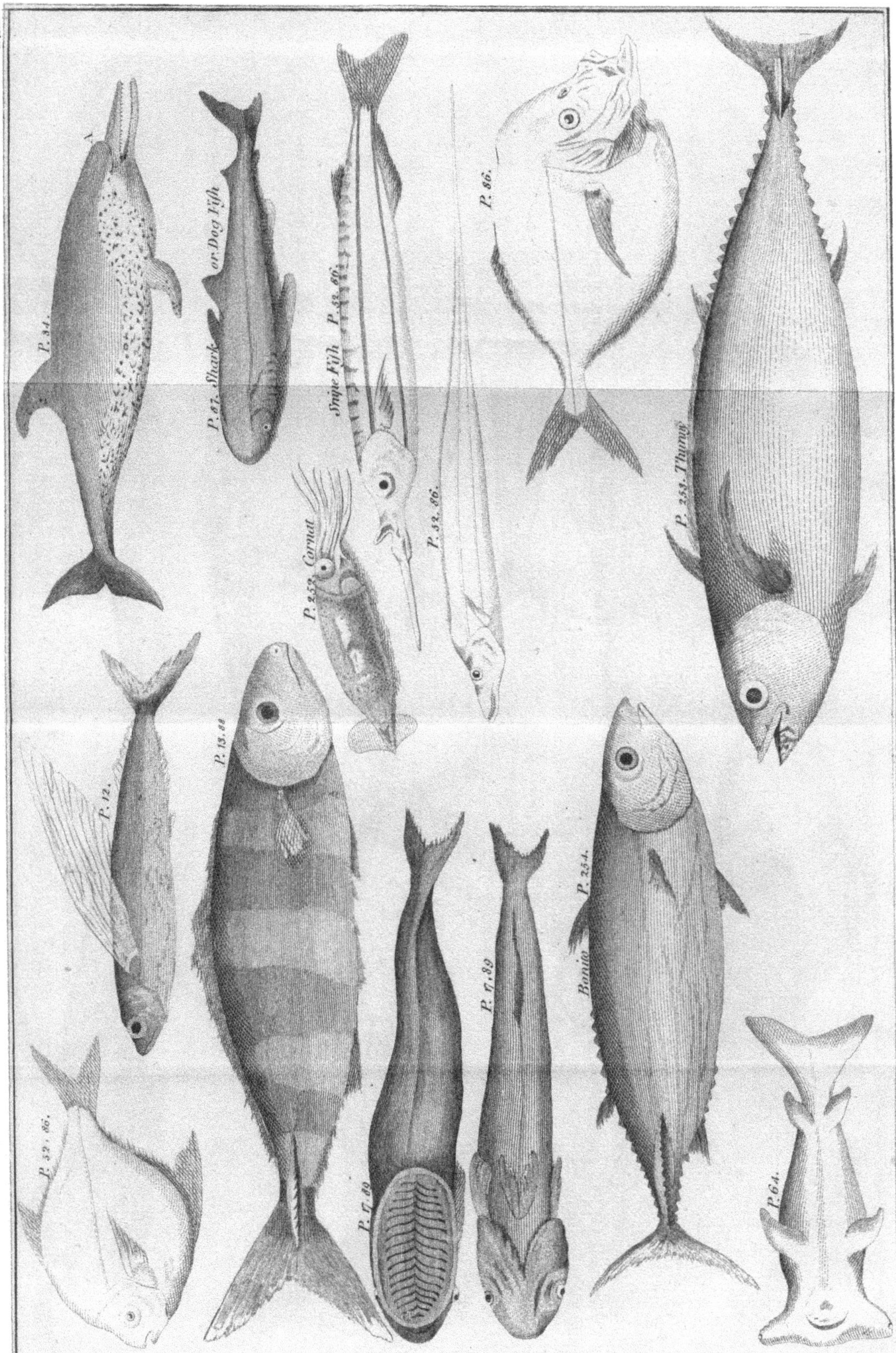

The material originally positioned here is too large for reproduction in this reissue. A PDF can be downloaded from the web address given on page iv of this book, by clicking on 'Resources Available'.

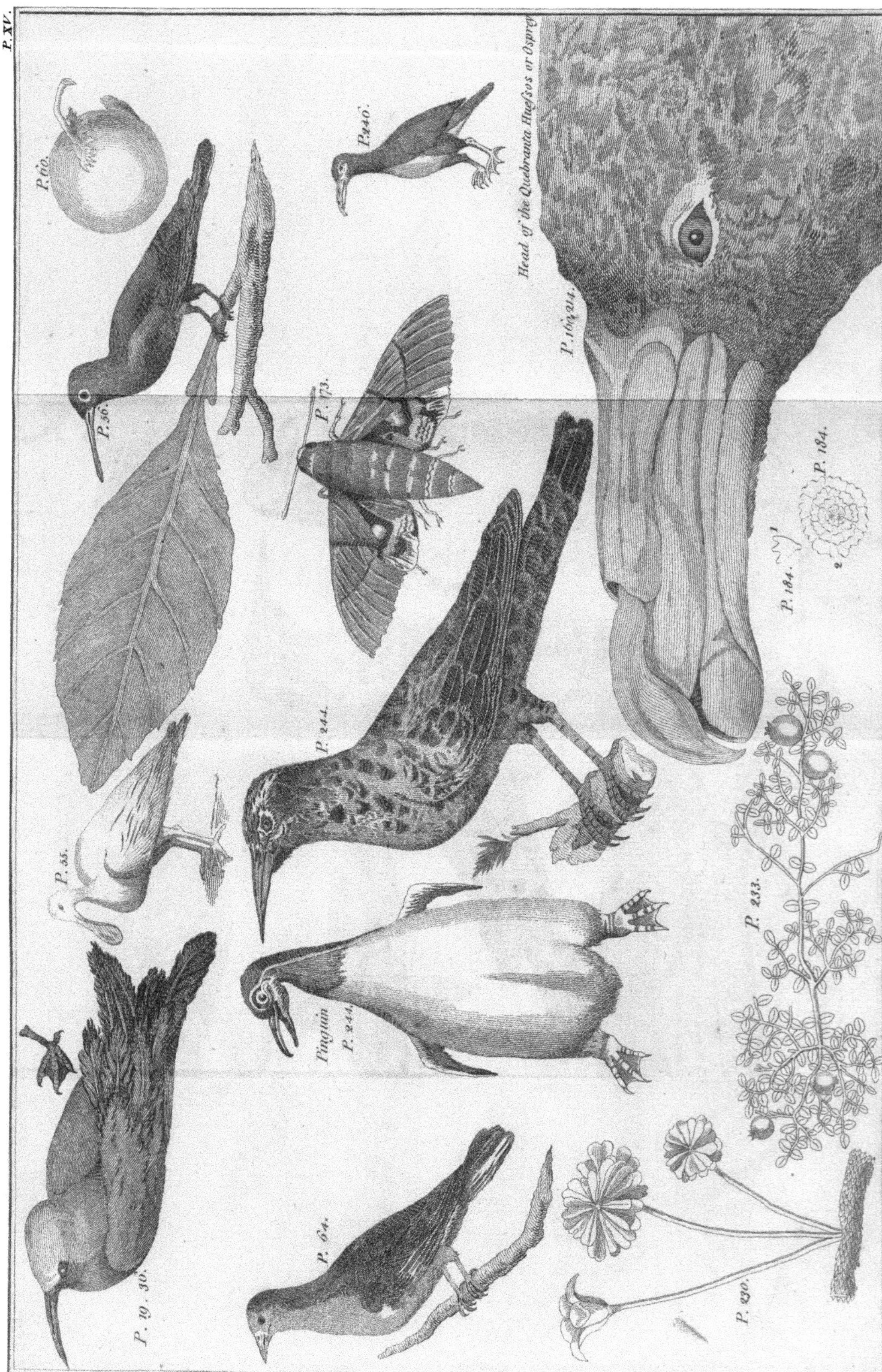

The material originally positioned here is too large for reproduction in this reissue. A PDF can be downloaded from the web address given on page iv of this book, by clicking on 'Resources Available'.

The material originally positioned here is too large for reproduction in this reissue. A PDF can be downloaded from the web address given on page iv of this book, by clicking on 'Resources Available'.

For EU product safety concerns, contact us at Calle de José Abascal, 56–1°,
28003 Madrid, Spain or eugpsr@cambridge.org.

www.ingramcontent.com/pod-product-compliance
Ingram Content Group UK Ltd.
Pitfield, Milton Keynes, MK11 3LW, UK
UKHW051451290926
15471UKWH00008B/1147
* 9 7 8 1 1 0 8 0 6 4 3 3 0 *